SpringerBriefs in Molecular Science

Chemistry of Foods

T0172003

Series Editor

Salvatore Parisi, Al-Balqa Applied University, Al-Salt, Jordan

The series Springer Briefs in Molecular Science: Chemistry of Foods presents compact topical volumes in the area of food chemistry. The series has a clear focus on the chemistry and chemical aspects of foods, topics such as the physics or biology of foods are not part of its scope. The Briefs volumes in the series aim at presenting chemical background information or an introduction and clear-cut overview on the chemistry related to specific topics in this area. Typical topics thus include:

- Compound classes in foods—their chemistry and properties with respect to the foods (e.g. sugars, proteins, fats, minerals, …)
- Contaminants and additives in foods—their chemistry and chemical transformations
- Chemical analysis and monitoring of foods
- Chemical transformations in foods, evolution and alterations of chemicals in foods, interactions between food and its packaging materials, chemical aspects of the food production processes
- Chemistry and the food industry—from safety protocols to modern food production

The treated subjects will particularly appeal to professionals and researchers concerned with food chemistry. Many volume topics address professionals and current problems in the food industry, but will also be interesting for readers generally concerned with the chemistry of foods. With the unique format and character of SpringerBriefs (50 to 125 pages), the volumes are compact and easily digestible. Briefs allow authors to present their ideas and readers to absorb them with minimal time investment. Briefs will be published as part of Springer's eBook collection, with millions of users worldwide. In addition, Briefs will be available for individual print and electronic purchase. Briefs are characterized by fast, global electronic dissemination, standard publishing contracts, easy-to-use manuscript preparation and formatting guidelines, and expedited production schedules.

Both solicited and unsolicited manuscripts focusing on food chemistry are considered for publication in this series. Submitted manuscripts will be reviewed and decided by the series editor, Dr. Salvatore Parisi.

To submit a proposal or request further information, please contact Tanja Weyandt, Publishing Editor, via tanja.weyandt@springer.com or Dr. Salvatore Parisi, Book Series Editor, via drparisi@inwind.it or drsalparisi5@gmail.com

More information about this series at http://www.springer.com/series/11853

Marco Fiorino · Caterina Barone ·
Michele Barone · Marco Mason ·
Arpan Bhagat

Quality Systems in the Food Industry

 Springer

Marco Fiorino
Studio Tecnico Fiorino
Siracusa, Italy

Michele Barone
Associazione "Componiamo il Futuro"
(CO.I.F.) Palermo
Palermo, Italy

Arpan Bhagat
Food Microbiologist
Argyle, TX, USA

Caterina Barone
Associazione "Componiamo il Futuro"
(CO.I.F.) Palermo
Palermo, Italy

Marco Mason
University of Cartagena
Cartagena, Colombia

ISSN 2191-5407 ISSN 2191-5415 (electronic)
SpringerBriefs in Molecular Science
ISSN 2199-689X ISSN 2199-7209 (electronic)
Chemistry of Foods
ISBN 978-3-030-22552-0 ISBN 978-3-030-22553-7 (eBook)
https://doi.org/10.1007/978-3-030-22553-7

© The Author(s), under exclusive license to Springer Nature Switzerland AG 2019, corrected publication 2019

This work is subject to copyright. All rights are solely and exclusively licensed by the Publisher, whether the whole or part of the material is concerned, specifically the rights of translation, reprinting, reuse of illustrations, recitation, broadcasting, reproduction on microfilms or in any other physical way, and transmission or information storage and retrieval, electronic adaptation, computer software, or by similar or dissimilar methodology now known or hereafter developed.

The use of general descriptive names, registered names, trademarks, service marks, etc. in this publication does not imply, even in the absence of a specific statement, that such names are exempt from the relevant protective laws and regulations and therefore free for general use.

The publisher, the authors and the editors are safe to assume that the advice and information in this book are believed to be true and accurate at the date of publication. Neither the publisher nor the authors or the editors give a warranty, expressed or implied, with respect to the material contained herein or for any errors or omissions that may have been made. The publisher remains neutral with regard to jurisdictional claims in published maps and institutional affiliations.

This Springer imprint is published by the registered company Springer Nature Switzerland AG
The registered company address is: Gewerbestrasse 11, 6330 Cham, Switzerland

The original version of the book was revised: correct affiliation of the book author "Arpan Bhagat" has now been updated. The erratum to the book is available at https://doi.org/10.1007/978-3-030-22553-7

Contents

Chapter 1
Chemical Additives for Foods. Impact of Food-Related Quality System Certifications on the Management of Working Flows

Abstract This chapter examines the role of food-oriented (or 'food-centric') quality system standards in the modern food and beverage industry. In general, quality schemes are based on the international norm ISO 9001 and the 'Hazard Analysis and Critical Control Points' approach. However, the chapter also introduces new and improved 'food quality' schemes and other safety-oriented and preventive approaches, currently outlining that a standardisation for international equivalence (while maintaining necessary flexibility and independence) is not always easy. This chapter discusses current issues concerning food-grade additives by the chemical and quality management viewpoints. This analysis is helpful because the reliability of management systems, including traceability, the design of working flows, and other concerns related to the use of food additives, implies an external and third-party assessment. As a result, the need for independent and reliable assessment has introduced third-party certification bodies in the world of food production.

Keywords Food additive · Flow chart · GSFS · IFS Food · Processing aid · Quality system · Traceability

Abbreviations

ATS	Addition and temporary storage
BRC	British Retail Consortium
CAS	Chemical Abstract Service
EU	European Union
FS	Final storage
T1	First corridor
FAW	Food additives warehouse
FSMA	Food Safety Modernization Act
FSSC	Food Safety System Certification
T4	Fourth corridor
GMO	Genetically modified organism

© The Author(s), under exclusive license to Springer Nature Switzerland AG 2019
M. Fiorino et al., *Quality Systems in the Food Industry*, Chemistry of Foods,
https://doi.org/10.1007/978-3-030-22553-7_1

1

GFSI	Global Food Safety Initiative
$[MO]_{additives}$	Global multi-origin of food additives
$[MO]_{food}$	Global multi-origin of the final product
GSFS	Global Standard for Food Safety
HACCP	Hazard Analysis and Critical Control Points
HARPC	Hazard Analysis and Risk-based Preventive Controls
IFS	International Featured Standard
ISO	International Organization for Standardization
MR	Mixing room
PO	Packaging operations
PMW	Packaging warehouse
PEA	Pre-entering area
RMW	Raw materials warehouse
RSPO	Roundtable on Sustainable Palm Oil
SQF	Safe Quality Food
T2	Second corridor
TS	Temporary storage area
T3	Third corridor
USA	United States of America

1.1 Chemical Additives for Food Productions. An Introduction

Food industries cannot avoid the use of minor (and necessary) ingredients in their formulations. This simple statement may be obvious when speaking of normal food science technology because the most part of all possible foods and beverages have always been obtained by means of the use of peculiar additives or processing aids (Ames et al. 1990; Broome and Hickey 1990; Deane and Hammond 1960; Fitzgerald and Buckley 1985; Ledford et al. 1966; Lück 1985; Nuñez et al. 1989; Okigbo et al. 1985; Oser 1985; Shehata et al. 1967; Woodroof 1966). As a simple and enlightening example, cheeses cannot be obtained from the original milk without the use of peculiar technologies using precipitating enzymes such as animal, vegetable or microbiological rennet, and/or the addition of certain acids or salts (citric acid and calcium chloride are well known in this broad ambit) (Abou-Zeid et al. 1983; Keller et al. 1974).

However, the average consumer of foods and beverages is always concerned when discussing food origin in a broad sense (Kendall et al. 2018; Zarling 2018). This behaviour has encouraged industries and researchers in the public and the private sectors with reference to new possible authentication studies and analytical methods (Fadini and Schnepel 1989; Forina et al. 1987; Szpylka et al. 2018a, b; Tsimidou et al. 1987; Williams 1985).

Actually, the analytical research in the chemical and microbiological ambits has always been carried out with reference to two distinct directions at least:

(a) The assessment of the real origin of foods and beverages intended as final products, and
(b) The evaluation of the origin of food and beverage ingredients as used in the food industry.

Apparently, these directions seem identical. On the other hand, it has to be considered that certain food and beverage products are mainly obtained from one raw material corresponding to the 85–95% at least of the original raw materials, while other products may contain more that 5–10 ingredients and additives. In relation to a simple example, cheese, two similar products may be realised in different ways when speaking of the number of introduced raw materials and other ingredients (Sagara et al. 1990; Shaw 1984; Johnson 1981; Torres and Chandan 1981; Trecker and Monckton 1990; Wargel et al. 1981):

(1) A common hard cheese from cow milk (intended use: grating) can be obtained with milk, rennet, salt (sodium chloride) and water. In this situation, milk may correspond to a minimum 95% of the total mass of raw materials, while salt may slightly exceed 1%, and water may arrive to 4%. The remaining ingredient—or a technological aid—is rennet, with a really negligible amount (<0.01%)
(2) Alternatively, a hard processed cheese for grating purposes may be also obtained by means of the use of different raw materials of milk or non-milk origin: water, rennet casein, various cheeses, milk butter, carbohydrates (potato starch, etc.) and other minoritary components including stabilisers, humectants agents, melting agents, etc.

By the viewpoint of average consumers, the difference between these products is related to two different problems:

(a) The definition of 'cheese' (real cheese is obtained from milk only with a minor presence of non-milk components, while processed or analogous cheeses are a mixture of different ingredients of mineral, vegetable, natural or synthetic origin)
(b) The definition of the origin of all components for the purchased product.

With reference to the exigency of average consumers, it has to be highlighted that the definition of food products may be different from the definition of ingredients used for the production. In other words (Mania et al. 2018a):

(1) The identification of a food product with a common name is linked to the natural or synthetic origin of ingredients (including the nature of original sources), the production method, the geographical origin of all raw materials and so on
(2) On the other hand, the origin of each raw material or ingredient (including food-grade chemical substances) can complicate the definition of the final food product. In fact, 'real' cheeses (or traditional foods such as olive oils or baked beans) are easily linked to the origin of a few ingredients, probably manufactured

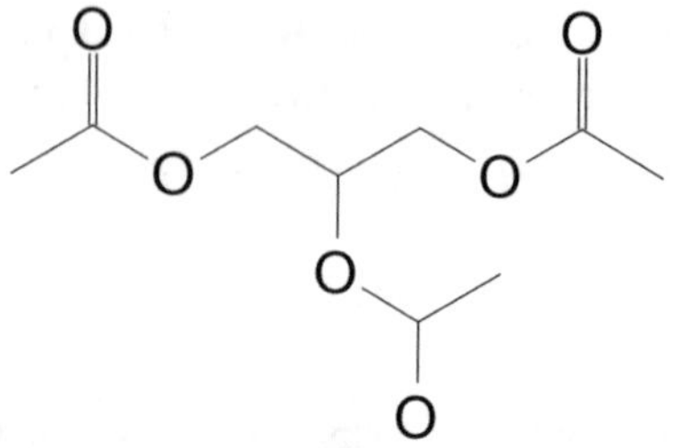

Fig. 1.1 Molecular structure of triacetin or glyceryl triacetate, a well-known additive in use in the food industry (also named E 1518). This compound—molecular formula: $C_9H_{14}O_6$, molecular weight 218.20 Da, Chemical Abstract Service (CAS) number 102-76-1—can be used in the food industry as a humectant agent

in the same region or country, and with a common or legally imposed production method (Delgado et al. 2017). On the other side, foods or beverages obtained by means of the use of a long list of ingredients should probably have a mixed or jeopardised origin.

An enlightening example may concern a well-known additive in use in the food industry: triacetin or glyceryl triacetate, named in the European Union (EU) as E 1518 (Bremus et al. 1983; Codex Alimentarius Commission 1995). This compound—molecular formula: $C_9H_{14}O_6$, molecular weight 218.20 Da, Chemical Abstract Service (CAS) number 102-76-1 (Fig. 1.1)—can be used in the food industry as a humectant agent, and other uses are known: antifungal, plasticising and flavouring properties are reported (Quinn and Ziolkowski 2015). By the chemical viewpoint, it is generally obtained from glycerol and acetic acid. Consequently, the related use of this additive in the modern world of food production may be notable enough, in spite of its predictably negligible amount in food formulations. When speaking of the origin of this compound, it is obtained by means of synthetic processes (non-natural origin) from glycerol and acetic acid.

In relation to the origin of pre-existing components, both glycerol and acetic acids may be different (animal or vegetable sources? various geographical origins?). As a simple result, the definition—or traceability exercise—applied to the situation of E 1518 (a minor component!) can be challenging enough, and give mixed answers, which could be questioned by average consumers.

Naturally, this compound is allowed in food production, but its origin may have some impact when speaking of the final definition and claims for the food: 'natural', '100% milk', '100% Italian' (or other country origins), 'vegetarian', 'vegan' and so on.

For these reasons at least, the use of food additives, processing aids and chemical substances in general should be preventively considered in the ambit of reliable and demonstrable traceability (Barbieri et al. 2014; Mania et al. 2018a, b, c). Generally, this ambit is covered also by quality certification systems for food products and beverages.

Table 1.1 A simple table showing allowed names for food additives, processing aids and food-grade chemical compounds in general in the EU ambit[2]

Authorised additives in the European Union: declared functions and related names			
Acid	Acidity regulator	Anti-caking agent	Anti-foaming agent
Antioxidant	Bulking agent	Colour	Emulsifier
Emulsifying salts	Firming agent	Flavour enhancer	Flour treatment agent
Foaming agent	Gelling agent	Glazing agent	Humectant
Modified starch	Preservative	Propellant gas	Raising agent
Sequestrant	Stabiliser	Sweetener	Thickener

Basically, a food or beverage product including one, two, three… 'n' chemical substances without a well-known name such as a normal food ingredient has to be evaluated one, two, three… 'n' times when speaking of the origin of all possible raw materials (Barbieri et al. 2014). As a result, the matter of authenticity is—first of all—a chemical matter.

Table 1.1 shows the list of allowed names for food additives, processing aids and food-grade chemical substances in general when speaking of the European legislation (European Parliament and Council 2011). This list is exhaustive enough, and it has to be used with the aim of giving a brief overview of problems food technologists and legal representatives working in food industries have to face nowadays. A total of 24 different functions are represented in Table 1.1. As a consequence, it may be inferred that the realisation of a complex food may easily involve at least 2–3 of these functions (and one or more chemical substances for each of these functions also).

In addition, there is a class of additives without mention in Table 1.1: the group of flavourings. In the European ambit, these substances are ruled by means of the Regulation (EC) No 1334/2008; also, it should be mentioned that two peculiar flavouring agents—caffeine and quinine—have to be declared not only with their function but also with the exact name (European Parliament and Council 2008).

Each function may be related to the chemical structure or classification of several food-grade chemical substances, as reported recently. Also, the common use and declaration of food additives includes the definition of chemical substances by means of a peculiar code, as requested in the European Union by the Regulation (EC) No 1333/2008 (Saltmarsh 2000). According to this system, four groups of food-grade chemical compounds can be identified (Laganà et al. 2017; Parisi 2017, 2018; Saltmarsh 2000):

(a) Additives used for protection (antioxidants, antimicrobial substances
(b) Colourant compounds
(c) Surrogates for sugars and sweetening agents
(d) Chemical compounds able to give a good and permanent structural behaviour and peculiar technological properties to the final food or beverage).

Anyway, the classification of different food additives is interesting only because food technologists may be able to reduce the complication derived from the origin of multiple substances by using only one (or two) different chemical substances in the same class. This behaviour corresponds to a precise technological strategy on the one side; however, each flavouring, thickener, sweetening substance and other compounds are generally prepared 'as they are named': their formulations require often the use of other substances!

As a simple result, should a peculiar food needs one, two, … 'n' different additives in the formulation, the following equation showing the 'global multi-origin of food additives' or $[MO]_{additives}$ of the complete mass of additives (with the exclusion of main raw materials) would be easily verified if each of these additives is also a carrier of one, two, … 'm' 'second-level' additives as shown in Eq. 1.1:

$$[MO]_{additives} = n \times m \tag{1.1}$$

Consequently, should this hypothetical food product have 'x' raw materials with 'y' origin features (geographical areas, natural sources, etc.), and 'n' additives/processing aids/food-grade chemical substances with 'm' possible origin features, the situation shown in Table 1.2 would be observed, where the global multi-origin of the final product—$[MO]_{food}$—can be obtained by means of Eq. 1.2. $[MO]_{food}$ corresponds to the global multi-origin of the final product assuming that $x = 1, y = 1$ and 'm' is $\leq$ 'n':

$$[MO]_{food} = x \times y + [MO]_{additives} = x \times y + n \times m \tag{1.2}$$

The complexity of the problem is determined by both the raw materials (generally representing 85–95% of the total mass of the food or beverage) and the used additives (minor components). Naturally, the importance of food-grade chemical is inversely proportional to their abundance when speaking of jeopardised origin…

This situation can be solved by means of traceability systems (Bosona and Gebresenbet 2013; Charlebois et al. 2014; Ene 2013; Lupien 2005; Mania et al. 2017, 2018a, b, c; Tian 2017). On the other side, the reliability of these systems implies an external and third-party assessment of the traceability procedures into food industries. The need for independent and reliable assessment has introduced third-party certification bodies in the world of food production.

1.1.1 Quality Certification Systems in the Food Industry

The world of foods and beverages has not initially been linked with the sector of quality management systems, although many companies and related representatives have followed the way of 'total quality' since 1990. The concept of quality assurance was initially related to food and non-food environments at the same time, with a peculiar fraction of customers seeking for accreditation in the non-food sectors. In

Table 1.2 Global multi-origin of food and beverage products $[MO]_{food}$ may be calculated by means of Eq. 1.2, where 'x' are raw materials, 'y' concern origin features and 'n' represent the number of additives/processing aids/food-grade chemicals with 'm' possible origin features

n	m	$[MO]_{food}$
1	1	3
2	1	4
2	2	5
2	1	4
2	2	5
2	3	6
3	1	5
3	2	6
3	3	7
4	1	6
4	2	7
4	3	8
4	4	9
5	1	7
5	2	8
5	3	9
5	4	10
5	5	11
6	1	8
6	2	9
6	3	10
6	4	11
6	5	12
6	6	13

Should $x = 1$, $y = 1$ and 'm' is $\leq$ 'n', $[MO]_{food}$ could be really challenging

other words, the possible advantages in terms of process improvement, higher profits and customer satisfaction were not an exclusive matter for food companies.

The birth of the ISO 9000 series of standards by the International Organization for Standardization (ISO) was a milestone when speaking of quality management systems. The above-mentioned norms aimed to set up recommendations and non-legally binging advices for companies and organisations seeking to ameliorate their results in terms of 'quality'. In brief, the goal of ISO was to help organisation in relation to the design, creation, delivery and continuous amelioration of supplied products and services by means of the demonstrable (reliable) prevention of 'non-conformities' (Brown et al. 1998; Buttle 1997; Dick 2000; Martínez Fuentes et al. 2000; Vloeberghs and Bellens 1996; Yung 1997).

However, the development of quality certification systems according to ISO 9000 norms has not been perceived positively by all possible institutions, including food and beverage operators. This behaviour has not been related to the complexity of quality systems or requirements, or to the comprehensible resistance against new and non-legally binding obligations. In addition, the existence of similar quality certifications based on the reliability of a quality management system located in a food company has generated some doubts when speaking of the dualism requested to the 'quality management responsible' (the subject responsible for quality improvement also directly chosen by the company direction).

On the contrary, many food and business operators have progressively questioned the general structure of quality management systems based on the 'Vision 2000' strategy (Conti 1999a, b; Renzi and Cappelli 2000). In detail, mass retailers above all have often considered the common structure of ISO 9000-based management systems as instruments without a strong connection with food quality, safety and integrity. For these and other reasons, the majority of food companies have voluntarily chosen (or have been forced) to comply with more 'food-centric' quality management systems (Stilo et al. 2009; Balasubramaniam et al. 2012; Bilska and Kowalski 2014; dos Santos Costa 2014; Filipović et al. 2008; Johnson 2014; Krieger and Schiefer 2007; Laux and Hurburgh 2012; SSICA, Certiquality, and Mobilpesca Surgelati SpA 2009; Sinha 2014; Trienekens and Zuurbier 2008; Van der Spiegel et al. 2003). This process towards the 'food quality management system' worldwide has progressively established a small number of related standards under the Global Food Safety Initiative (GFSI). At present, the most known certification standards in the broad ambit of food and packaging producers at least include (BRC 2018a; FSSC 2017; IFS 2017; ISO 2018; RSPO 2014; SQF 2017):

- The Global Standard for Food Safety (GSFS) by the British Retail Consortium (BRC)
- The International Featured Standard (IFS) Food, by the IFS
- The ISO 22000 norm, by the ISO
- The Food Safety System Certification (FSSC) 22000
- The Safe Quality Food (SQF) by the SQF Institute, USA
- The Roundtable on Sustainable Palm Oil (RSPO) certification scheme, by the homonym organisation.

In general, these systems have a common point: the partial or total adherence of the chosen procedures and instructions to basic principles of the Hazard Analysis and Critical Control Points (HACCP) assessment, although another US-related approach—the Hazard Analysis and Risk-based Preventive Controls (HARPC)—has to be considered before if its introduction some years ago (Bai et al. 2007; Baines 1999; Channaiah et al. 2017; Davidson et al. 2017; Grover et al. 2015; Parisi 2009, 2012, 2016; Gurnari 2015; Schulze et al. 2008; Stilo et al. 2009). Clearly, the basis of similar systems is always the 'old' ISO 9000 approach, but differences between food- and non-food environment have justified dissimilar strategies in response to the most part of consumer's worries (Askew 2018; Baş et al. 2007; Belik et al. 2001;

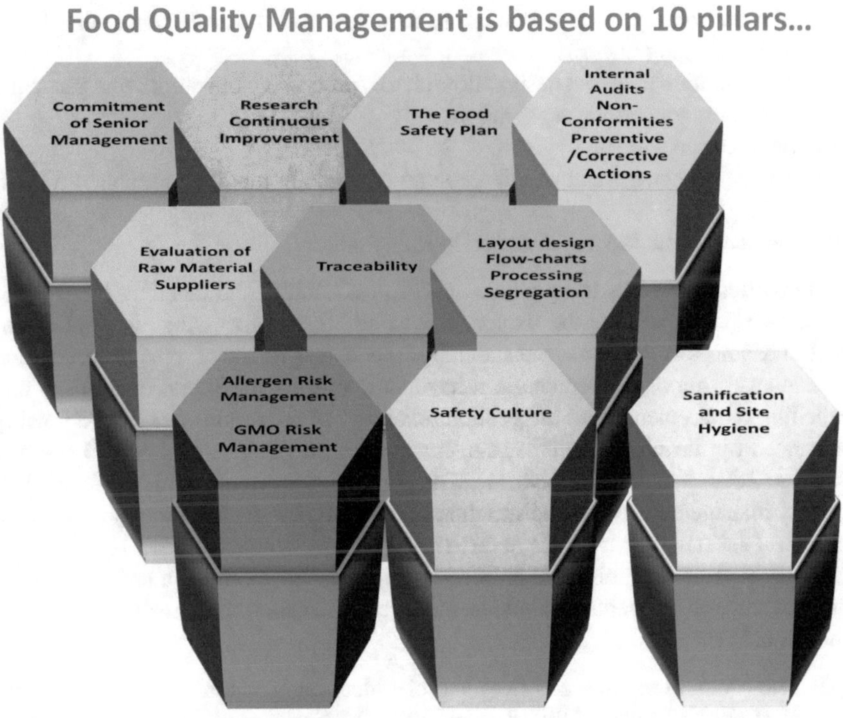

Food Quality Management is based on 10 pillars...

Fig. 1.2 Structure of quality management strategies in the food and beverage ambit is based on ten main pillars

Botonaki et al. 2006; Fulponi 2006; Knaflewska and Pospiech 2007; McMeekin et al. 2006; Wu et al. 2011).

In practice, the structure of these—and other—possible strategies when speaking of food quality management is based on the following pillars (Fig. 1.2) (Akkerman et al. 2010; Barone et al. 2015; BRC 2018b; Brunazzi et al. 2014; Escanciano and Santos-Vijande 2014; Fulponi et al. 2006; Mukundan 2005; Parisi and Luo 2018; Singla et al. 2018; Steinka et al. 2017; Zaccheo et al. 2017):

(a) The commitment of senior management
(b) The trend and research for continuous amelioration
(c) The elaboration, development and improvement of the 'food safety plan' (also intended according to the US Food Safety Modernization Act (FSMA)
(d) The performance of the organisation when speaking of internal audits, non-conformities and preventive/corrective actions
(e) The evaluation and re-assessment of raw material suppliers. There is no difference between edible materials and 'accessory' items such as packaging materials and objects

(f) The traceability, as legally binding requisite, with controls concerning labels and packaging integrity
(g) The attention to layout design, flow charts, processing operations and particular requisites concerning segregation
(h) Sanification and site hygiene
(i) The safe management of allergen and genetically modified organism (GMO) risks
(j) Dedicated training and attention to safety culture.

In particular, two of these pillars concern traceability and labelling procedures: these two points include the matter of food additives, processing aids and chemical substances in general. Substantially, the management of raw materials and 'secondary' (minor) ingredients can have an important impact on traceability and labelling requirements, and the general reliability of food industries can be notably enhanced by means of quality management systems (Coff et al. 2008; Opara and Mazaud 2001; Wognum et al. 2011). The real question should be: How should any quality management system be structured if traceability and labelling of food-grade chemical substances have to be assured?

By a general viewpoint, the problem of ingredients—both main and minor food and beverage components—should be managed with care when considering four key factors at least:

(1) The problem of food additives and chemical compounds in general (Asensio et al. 2008; Lindsay 2007; Lupien 2005; McEntire et al. 2010; Singhal et al. 1997): traceability, authenticity, position in the flow chart of a process
(2) The contrast to intentional adulteration (Fennema et al. 2017; Johnson 2014; Moore et al. 2012): fraud-related risk assessment and related mitigation strategies, supplier evaluation and continuous surveillance, new or improved analytical strategies, and economic studies and researches
(3) The examination of technical data sheets in the food and beverage industries by the viewpoint of external auditors (Crossland 1997; Panisello and Quantick 2001)
(4) The risk of declared and undeclared allergens BRC I 2017a, b; Nikoleiski 2015; Stein 2015) with a focus on 'masked' risk carriers (lubricants used for food processing equipments).

Actually, the management of foods and beverages by a global perspective is surely more complex than the above-shown list. However, the simple management of these points may be difficult enough: consequently, the aim of this book has been to give an overview of the current state of the art in these areas.

The first of these key factors is examined in this chapter, while:

(a) The contrast to intentional adulteration is discussed in Chap. 2
(b) The examination of technical data sheets in the food and beverage industries by the viewpoint of external auditors is examined in Chap. 3

(c) The risk of declared and undeclared allergens with a focus on 'masked' risk carriers such as lubricants used for food processing equipments is evaluated in Chap. 4.

As a premise, and in connection with main 'pillars' of food quality management systems (Fig. 1.2), it has to be considered that:

(1) The discussion concerning food additives and chemical compounds in a quality management ambit should at least include considerations related to: the commitment of senior management; the 'food safety plan'; the performance of the organisation when speaking of internal audits, non-conformities and preventive/corrective actions; the evaluation and re-assessment of raw material suppliers; traceability; flow charts, processing operations, and particular requisites concerning segregation; and sanification procedures

(2) The contrast to intentional adulteration (Chap. 2) concerns the following quality management areas: the commitment of senior management; the trend and research for continuous amelioration; the 'food safety plan' (risk assessment and dedicated procedures); the performance of the organisation (internal audits, non-conformities and preventive/corrective actions); the evaluation and re-assessment of raw material suppliers; traceability and labelling requirements; processing operations and related flow charts, with peculiar attention to site design; peculiar segregation and sanification procedures; training and attention to safety culture

(3) The examination of technical data sheets in the food and beverage industries (Chap. 3) has to consider at least: the 'food safety plan'; the performance of the organisation (internal audits, non-conformities and preventive/corrective actions); the evaluation and re-assessment of raw material suppliers; traceability; the safe management of allergen and GMO risks

(4) Finally, allergen risks and the use of peculiar lubricants (Chap. 4) should concern the following areas: the commitment of senior management; the trend and research for continuous amelioration; the 'food safety plan'; the performance of the organisation (internal audits, non-conformities and preventive/corrective actions); the evaluation and re-assessment of raw material suppliers (including accessory materials such as lubricants); traceability, layout design, flow charts, processing operations and particular requisites concerning segregation; sanification and site hygiene; and dedicated training and attention to safety culture.

Several of the above-mentioned areas are discussed in detail, while other points are briefly discussed because of their non-chemical adherence to the related factor (food additives and chemical compounds; food frauds; technical data sheets; and allergen risks).

1.1.2   How Can Food Additives Be Managed Correctly in a Quality-Oriented System? Main Requirements

The Global Standard for Food Safety, Issue 8, mentions 'additives' many times, and the importance of these minor ingredients is considered in many ambits of the quality management system (BRC 2018b). The same thing is observed in the IFS Food, version 6.1 (IFS 2017), and in other quality standards (Sect. 1.2).

First of all, the position of entering additives and chemical compounds in the process has to be clearly defined when speaking of flow diagrams, as required by GSFS (clause 2.6.1) and also the Codex Alimentarius Commission (BRC 2018b). More specifically, food additives have to be incorporated in the flow diagram when used. A critical point in this ambit is also related to the updated information concerning flow diagrams, because possible ameliorations of a food processing operations may require—or may exclude—the use of a particular food ingredient, depending on its function (Sect. 1.1). Consequently, information correlated to the presence or absences of a raw material have to be constantly updated. A similar request is implicitly stated in IFS Food, version 6.1: food processing plans should describe clearly the flow of all materials and products entering and exiting the process, also including waste, operators and water (clause 4.8.1). Interestingly, the requirement also cites the existence of a site map concerning also the design of flow charts (because of the obvious link between layout design and food processing design).

A really challenging point concerning the use of food additives and food-grade chemical substances in general is tacitly cited by GSFS in the clause 5.3.1 when speaking of raw material assessment for possible allergen contamination (BRC 2018b). Actually, the presence of food additives would not immediately linked to the control of raw materials in relation to possible allergen contamination. On the other side, it is recommended that raw material features (with their inner composition) are preliminarily discussed and agreed with raw material suppliers. Because of the possible and often verifiable presence of different chemical substances and mixtures (flavouring agents, processing aids, carriers, etc.), it has to be noted that the assessment and agreement of raw materials concerns also their previous history in terms of production. The presence of food additives should be considered by food producers not only in their process, but also in all pre-existing processes out of their surveillance (possible dangers: deliberate presence or potential cross-contamination). The basic example in Sect. 1.1 (Eqs. 1.1 and 1.2) concerns only the multi-variegated origin of a food product; however, this example highlights the role—on a numerical level—of all possible raw materials, including negligible amounts of additives, and their previous history. In relation to IFS Food, clause 4.20.1, the requirement is tacitly present when speaking of raw materials containing allergens.

Another good point when speaking of management of food additives and food-grade chemical substances in general concerns dedicated storage structures for all materials, including processing aids and additives, as requested by IFS Food, clause 4.14.4 (IFS 2017). Interestingly:

(a) The management of storage facilities should consider technical data sheets and related recommendations for food additives (Chap. 3)
(b) The packaging materials used for food additives have to be considered when speaking of safe storage
(c) Sanitisation procedures have to consider the presence of food additives on site because of possible packaging damages caused by improper sanitisation, and/or because of intrinsic features of some additive (e.g. hygroscopic food-grade chemical compounds should be adequately stored, protected against humidity and reprotected when speaking of opened packages)
(d) Training is needed when speaking of safe storage for operators? Clearly, the answer should be positive.

In relation to IFS Food, version 6.1, it should be also noted that a particular recommendation for auditors concerns the critical closure 4.2.1.2 concerning raw materials and their specifications. More in detail, auditors are requested to clearly identify the name of specifications for checked materials (including additives also). This observation appears obvious; however, there are many situations concerning peculiar additives and processing aids, or mixtures of them (with carriers, flavourings, etc.) and named with commercial names instead of their specific name. As a result, it may be assumed that a generic product defined briefly as 'potato starch preparation' is intended as 'potato starch' only, while the formulation includes also different additives. The identification of specifications means (i) the use of technical data sheets and (ii) the definition of the used preparation 'as it is' instead of a common definition for similarity… Another mention is made in IFS Food, version 6.1 (clause 4.19.1) in relation to each possible chemical substance (flavourings, additives and so on) which could be derived from GMO, even if labelling norms do not require their mention (IFS 2017).

Finally, the problem of traceability concerns all raw materials, including food additives also (the exact definition does not matter). According to GSFS Issue 9, clause 3.9.2, all possible ingredients (and food contact packaging materials also) are 'raw materials; for this reason, they have to be traced and mentioned even if their mention on labels is not needed (BRC 2018b). Consequently, traceability systems have to consider always mentioned and not-mentioned ingredients for a food or beverage product!

1.1.3   Working Flows and Flow Charts in the Food Industry. Reliable Accounts

Ideally, the production in food and beverage industries would be a dynamical process along a theoretically one-way direction. This simple declaration is easily questionable in the real world because of the following elements (BRC 2018b; Mania et al. 2017, 2018a, b, c; Parisi 2005):

(a) There are more than one possible exiting materials from the process for one, two, … 'n' entering raw material, additive and/or packaging material: the final and desired products, various sub-products and discarder pieces (and packaging discards)

(b) In addition, various processes use in some step fluid carriers such as simple water. As a result, should this fluid serve only as a carrier (without absorption or incorporation in the final product, or in one of by-products), the exiting flow of materials would also include the fluid itself

(c) Depending on processes, certain materials can be elaborated with the intra-processing recycling of discarded products (on condition that safety, legality, integrity and quality of the product remain guaranteed). The recycle or re-use may involve discarded pieces during the process itself at the same date, or in different dates.

On these bases, there is not a one-way direction, from raw materials to finished products, but a fragmented process including one or more entering materials and one or more exiting materials, in different points of the flow.

The 'geographical' position of entering and exiting materials in the process is critical, also in connection with the creation of logical and reliable flow charts (Alli 2003; Ibarz and Barbosa-Cánovas 2002; Sun and Ockerman 2005; Tanner 2000). In other words, the logical flow of processing operations should aim at the reduction of contamination episodes by means of the limitation of all possible risks, following the broad FSMA (or HARPC) approach (Alli 2003; Grover et al. 2016; Koenig 2011; Kowalska 2018; Johnson 2014).

This result, extremely difficult to achieve in a chaotic environment, has been obtained by means of the classification of all possible risks, their localisation into the site (and off-site also) and the management of these dangers by means of dedicated countermeasures. The first—or one of the main—of these measures should be the definition of one way only when speaking of processing directions: from raw materials to finished products, without deviations in the opposite direction (Luning et al. 2006).

Moreover, the spatial localisation of steps composing the entire flow diagram is critical. In general, and in relation to all possible sites realised years ago, the most frequent situation concerns food plants without a well-defined layout, if observed with the eyes of the food technologist (Amjadi and Hussain 2005; Anonymous 1977; Moerman and Wouters 2016; UNECE 1981; Waldron 2009). In other words, and with specific relation to entering raw materials (including food additives), the best strategy should be the limitation of entering possibilities: ideally, one single enter point would be desirable, while more than one exit points may exist because of the occurrence of one or more sub-products (and fluids) (Bindi et al. 2009; Wanniarachchi et al. 2016). This objective can be obtained by means of logical and predesigned layout projects (Daf and Zanwar 2013; Drira et al. 2007; Holah 2003; Shahin and Poormostafa 2011).

In addition, a new and emerging menace should be considered: the possibility of deliberate and malicious acts against food companies, also intended industrial sabotage (Kanai et al. 2015; Van Donk and Gaalman 2004; Wanniarachchi et al.

2016). One of the most known possibilities, when speaking of deliberate attacks to food companies, concerns sabotage of processing operations. As a consequence, raw materials should really be monitored on site in relation to the following possibilities at least:

(a) Deliberate substitution of one raw material with another one
(b) Malicious contamination of one raw material with another one
(c) Manumission of the processing flow by means of the augment of one raw material if compared with normal formulations, with labelling implications also
(d) Manumission of the processing flow by means of the diminution or the complete elimination of one raw material if compared with normal formulations, with labelling implications also.

With reference to raw materials, contamination may be prevented but dangers are real; on the other side, the augment or the diminution/elimination from the operation formula is difficult enough because of the predictably notable amount of main raw materials in the product.

On the other side, food additives and food-grade chemical substances in general are negligible in terms of formulation amounts: these compounds may globally range from 0.5 to 5.0% of the total formula in many situations. As a result, food defence plans have to be considered in relation to food additives in general because malicious attacks in this ambit might be unnoticed.

Anyway, processing flow diagrams should tend to the limitation of all possible entering accesses when speaking of raw materials. A simplified example can be shown in Figs. 1.3 and 1.4.

In brief, Fig. 1.3 shows a simplified food production site subdivided in the following areas (all areas, rooms and corridors are identified with specific identification codes):

(a) Raw materials (warehouse), code: RMW
(b) Food additives (warehouse), code: FAW
(c) A pre-entering area (needed for opening operations and safety inspections before transport to food production areas), code: PEA
(d) A first corridor from the pre-entering area to a dedicated mixing room, code: T1
(e) The mixing room (where raw materials and additives are used together and processes), code: MR
(f) The second corridor between mixing room and a temporary storage area, code: T2
(g) The temporary storage area, code: TS
(h) The third corridor between the temporary storage area and the 'addition and temporary storage' room, code: T3
(i) The 'addition and temporary storage' room where the process requires the addition of a second food additive and immediately another temporary storage, code: ATS

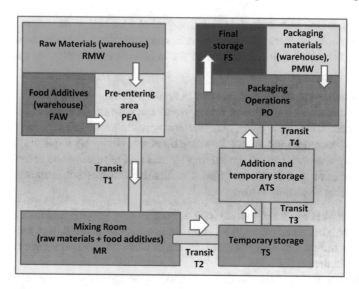

Fig. 1.3 A food production site is subdivided in different areas and rooms, depending on the specific function. This picture shows a simplified site subdivided as follows (all areas, rooms and corridors are identified with specific identification codes): raw materials (warehouse), code: RMW; food additives (warehouse), code: FAW; the pre-entering area, code: PEA; a corridor from the pre-entering area to the dedicated mixing room, code: T1; the mixing room, code: MR; another corridor between mixing room and a temporary storage area, code: T2; the temporary storage area, code: TS; another corridor between the temporary storage area and the 'addition and temporary storage' room, code: T3; the 'addition and temporary storage' room, code: ATS; another corridor between the 'addition and temporary storage' room and the 'packaging operations' area, code: T4; another warehouse containing packaging materials only, code: PMW; the 'packaging operations' area, code: PO; and the 'final storage' area, code: FS

(j)   The fourth corridor between the 'addition and temporary storage' room and the 'packaging operations' area, code: T4

(k)   The packaging warehouse containing packaging materials only (these materials are transported to the 'packaging operations' area), code: PMW

(l)   The 'packaging operations' area, code: PO

(m)   The 'final storage' area, code: FS.

Figure 1.4 shows a simplified flow diagram for the example process, taking into account Fig. 1.3 and specific codes. The simplified flow chart shows only mixing, addition and temporary storage, and packaging operations because these procedures concern only entering materials, and one exiting product (in the PO area). Each ingredient entry is highlighted with a peculiar ENTRY symbol. Our example shows:

(1)   Three raw materials and one food additives entering the process in MR

(2)   One additional food additive entering the process in ATS

(3)   Three packaging materials entering the process in PO.

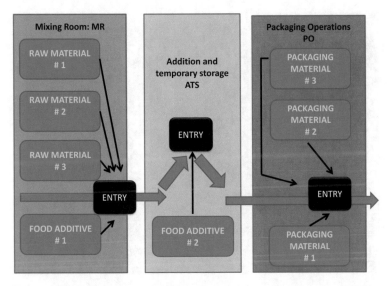

Fig. 1.4 A simplified flow diagram for the example process, taking into account Fig. 1.3 and specific codes. The simplified flow chart shows only mixing, addition and temporary storage, and packaging operations because these procedures concern only entering materials and one exiting product (in the PO area). Each ingredient entry is highlighted with a peculiar ENTRY symbol

The simplified process shows only three raw materials, two additives and three packaging materials, but the position in the process needs three entry places. On the one side, this situation can explain well reasons for limiting the number of possible entry places in the process. Unfortunately, certain processes cannot be modified in this direction, forcing food industries to search for the most important limitation of entering possibilities in terms of ingredients and entry areas. It has to be noted that malicious food attacks can easily occur in the three entry areas at least. The higher the number of raw materials and food additives, the higher the possible risks concerning HACCP/HARPC risks, identity loss and malicious attacks. At the same time, the higher the number of entry areas, the higher the possibilities for sabotages, identity loss and food safety.

Moreover, the simplified process does not show a two-way process, but an 'ideal' one-way process (from raw materials and ingredients including packaging materials to the final product). Should the process have one deviation from the ideality—for example, a portion of processed intermediates in ATS (Fig. 1.5) may exit temporarily from the process and be recycled immediately in MR as a secondary raw material or 'off-line' product (Mania et al. 2018b, c, d)—some possible formulation error could occur because the initial formulation does not contemplate secondary raw materials. In this situation, the number of entry nodes is not modified; however, should the possibility of recycling occur in different areas (more than one single node, differently from our example), the risk of non-conformities and malicious attacks would increase dramatically. A simple example (Fig. 1.6) showing four different entry areas or nodes

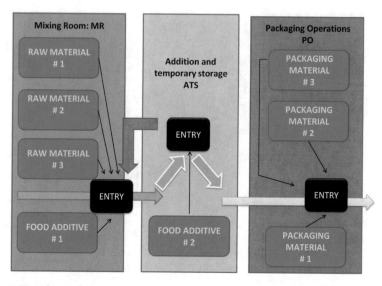

Fig. 1.5 Simplified process shown in Fig. 1.4 does not show a two-way process, but an 'ideal' one-way process (from raw materials and ingredients including packaging materials to the final product). Should a portion of processed intermediates in ATS exit temporarily from the process and be recycled immediately in MR as a secondary raw material or 'off-line' product, some possible formulation error could occur because the initial formulation does not contemplate secondary raw materials. In this situation, the number of entry nodes would not be modified but the risk of non-conformities and malicious attacks would increase dramatically

can highlight the complexity of the problem in food industries: the first of these nodes does not contemplate entering 'off-line' raw materials (containing used food additives), while entries 2, 3 and 4 (the second entry is really challenging) may receive secondary raw materials or intermediates. It has to be noted that entry 2 can receive:

(a) One intermediate from entry 1 (direction: start → end)
(b) One secondary intermediate (direction: start ← end) from entry 4

and send:

(c) One intermediate to entry 3 (direction: start → end)
(d) One secondary intermediate to 'start' (direction: start ← end).

In other words, the complexity of the whole network may be calculated in terms of number of exiting and entering 'movements' or 'connections'—the displacement of one intermediate food or material from one node to another node, without mention of the direction—into the network.[3] However, complexity of networks does not

---

[3]These and other information are discussed in the 'Food Traceability' lecture for the Faculty of Agricultural Technology, Al-Balqa Applied University, Al-Salt, Jordan, by Salvatore Parisi, first course 2018–2019.

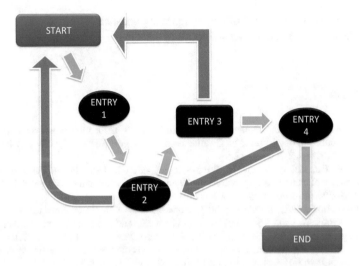

Fig. 1.6 A simple example showing four different entry areas or nodes can highlight the complexity of the problem in food industries: the first of these nodes does not contemplate entering 'off-line' raw materials while entries 2, 3 and 4 may receive secondary raw materials or intermediates. This network has eight connections while the total expected number of connections may be. In this situation, the % complexity of the network is only '8/30 × 100' = 27%

necessarily correspond to the complexity of accesses. In the first example (Fig. 1.3), this number is small (four connections) while the total expected number of connections may be 'three (nodes) × two (theoretical possible connections = number of nodes—1)' = 6 movements. As a result, the percentage of possible nodes—also considered as the 'complexity' of the network—would be '4/6 × 100' = 67%. In relation to Fig. 1.5, the number of exiting and entering connections into the network (five) demonstrates that the % complexity is augmented if compared with Fig. 1.3: '5/6 × 100' = 83%. Finally, the basic network shown in Fig. 1.6 has eight connections while the total expected number of connections may be 'six (nodes) × five (theoretical possible movements)' = 30 movements. In this situation, the % complexity of the network is only '8/30 × 100' = 27%.

On the other side, the access number corresponds to the entire number of all raw materials, foods additives, secondary raw materials and packaging materials which can enter into an 'entry node'. The network complexity cannot give an idea of risks correlated to entering and exiting materials. The above-mentioned calculations should not only consider the number of access nodes (entries) and all possible interconnections between nodes, but also the possibility that one interconnection can concern one, two, … 'n' possible entering and exiting materials! As a consequence, and as a basic conclusion, the work of food technologists on logic networks should initially aim at (a) the reduction of entry or access nodes, (b) the reduction of connections for each node and finally (c) at the limitation of entering and exiting ingredients in the flow diagram. The last point could be considered first of all, but it can be often placed in the end of this design work because of limited options allowed when speaking of certain processed foods.

References

Abou-Zeid AZA, Baghlaf AO, Khan JA, Makhashin SS (1983) Utilization of date seeds and cheese whey in production of citric acid by Candida lipolytica. Agric Wastes 8(3):131–142. https://doi.org/10.1016/0141-4607(83)90112-9

Akkerman R, Farahani P, Grunow M (2010) Quality, safety and sustainability in food distribution: a review of quantitative operations management approaches and challenges. OR Spectrum 32(4):863–904. https://doi.org/10.1007/s00291-010-0223-2

Alli I (2003) Food quality assurance: principles and practices. CRC Press, Boca Raton

Ames BN, Profet M, Gold LS (1990) Nature's chemicals and synthetic chemicals: comparative toxicology. Proc Natl Acad Sci 87(19):7782–7786. https://doi.org/10.1073/pnas.87.19.7782

Amjadi K, Hussain K (2005) Integrating food hygiene into quantity food production systems. Nutr Food Sci 35(3):169–183. https://doi.org/10.1108/00346650510594921

Anonymous (1977) Chilton's food engineering, vol 49. Chilton Company, Radnor

Asensio L, González I, García T, Martín R (2008) Determination of food authenticity by enzyme-linked immunosorbent assay (ELISA). Food Control 19(1):1–8. https://doi.org/10.1016/j.foodcont.2007.02.010

Askew K (2018) Sugar replaces price as consumers' top food worry. FoodNavigator.com. Available https://www.foodnavigator.com/Article/2018/08/20/Sugar-replaces-price-as-consumers-top-food-worry. Accessed 08 Apr 2019

Bai L, Ma C, Gong S, Yang Y (2007) Food safety assurance systems in China. Food Control 18(5):480–484. https://doi.org/10.1016/j.foodcont.2005.12.005

Baines RN (1999) Environmental & animal safety dimensions to developing food safety & quality assurance initiatives in the United Kingdom. Market rights and equity: food and agriculture standards in a shrinking world. Institute for Food & Agricultural Standards, Michigan

Balasubramaniam S, Jeeva JC, Ashaletha S (2012) Adoption of quality management practices in seafood processing sector in cochin region. Fish Technol 49:80–86. Available http://drs.cift.res.in/handle/123456789/686. Accessed 11 Apr 2019

Barbieri G, Barone C, Bhagat A, Caruso G, Conley ZR, Parisi S (2014) The influence of chemistry on new foods and traditional products. Springer briefs in chemistry of foods. Springer International Publishing, Heidelberg

Barone C, Bolzoni L, Caruso G, Montanari A, Parisi S, Steinka I (2015) Food packaging hygiene. Springer Briefs in Chemistry of Foods. Springer International Publishing, Cham. https://doi.org/10.1007/978-3-319-14827-4

Baş M, Yüksel M, Çavuşoğlu T (2007) Difficulties and barriers for the implementing of HACCP and food safety systems in food businesses in Turkey. Food Control 18(2):124–130. https://doi.org/10.1016/j.foodcont.2005.09.002

Belik W, dos Santos RR, Green R (2001) Food retailing, quality signals and the customer defence. In: Proceedings of the world food and agribusiness symposium of the International Food and Agrobusiness Management Association (IAMA), Sidney. Available http://www.agrifood.info/Agrifood/members/Congress/Congress2001Papers/Symposium/Belik.pdf. Accessed 10 Apr 2019

Bilska A, Kowalski R (2014) Food quality and safety management. Sci J Log 10(3):351–361

Bindi F, Manzini R, Pareschi A, Regattieri A (2009) Similarity-based storage allocation rules in an order picking system: an application to the food service industry. Int J Logist Res Appl 12(4):233–247. https://doi.org/10.1080/13675560903075943

Bosona T, Gebresenbet G (2013) Food traceability as an integral part of logistics management in food and agricultural supply chain. Food Control 33(1):32–48. https://doi.org/10.1016/j.foodcont.2013.02.004

Botonaki A, Polymeros K, Tsakiridou E, Mattas K (2006) The role of food quality certification on consumers' food choices. Brit Food J 108(2):77–90. https://doi.org/10.1108/00070700610644906

BRC (2017a) Global standard food safety, issue 7—An introduction to best-practice lubrication procedures in the food industry. British Retail Consortium (BRC) Global Standards, London, www.brcglobalstandards.com. Available https://www.brcgs.com/media/165616/brc-lubrication-guide-screen.pdf. Accessed 10 Apr 2019

BRC (2017b) An introduction to best-practice lubrication procedures in the food industry, issue 1, April 2017. British Retail Consortium (BRC) Global Standards, London, www.brcglobalstandards.com

BRC (2018a) Global standard food safety, issue 8. British Retail Consortium (BRC) Global Standards, London, www.brcglobalstandards.com

BRC (2018b) Global standard food safety, issue 8. Interpretation Guideline. British Retail Consortium (BRC) Global Standards, London, www.brcglobalstandards.com

Bremus N, Dieckelmann G, Jeromin L, Rupilius W, Schutt H (1983) Process for the continuous production of triacetin. US Patent Application 4,381,407, 26 Apr 1983

Broome MC, Hickey MW (1990) Comparison of fermentation produced chymosin and calf rennet in Cheddar cheese. Aust J Dairy Technol 45(2):53–59

Brown A, Van Der Wiele T, Loughton K (1998) Smaller enterprises' experiences with ISO 9000. Int J Qual Reliab Manag 15(3):273–285. https://doi.org/10.1108/02656719810198935

Brunazzi G, Parisi S, Pereno A (2014) The importance of packaging design for the chemistry of food products. Springer Briefs in Chemistry of Foods. Springer International Publishing, Cham. https://doi.org/10.1007/978-3-319-08452-7

Buttle F (1997) ISO 9000: marketing motivations and benefits. Int J Qual Reliab Manag 14(9):936–947. https://doi.org/10.1108/02656719710186867

Channaiah LH, Michael M, Acuff JC, Phebus RK, Thippareddi H, Olewnik M, Milliken G (2017) Validation of the baking process as a kill-step for controlling Salmonella in muffins. Int J Food Microbiol 250:1–6. https://doi.org/10.1016/j.ijfoodmicro.2017.03.007

Charlebois S, Sterling B, Haratifar S, Naing SK (2014) Comparison of global food traceability regulations and requirements. Compr Rev Food Sci Food Saf 13(5):1104–1123. https://doi.org/10.1111/1541-4337.12101

Codex Alimentarius Commission (1995) General standard for food additives—CODEX STAN 192–1995, last revision 2018. The Food and Agriculture Organization of the United Nations (FAO), Rome, and the World Health Organization (WHO), Geneva. Available http://www.fao.org/gsfaonline/index.html. Accessed 08 Apr 2019

Coff C, Korthals M, Barling D (2008) Ethical traceability and informed food choice. In: Coff C, Barling D, Korthals M, Nielsen T (eds) Ethical traceability and communicating food, the international library of environmental, agricultural and food ethics, vol 15, pp 1–18. Springer Science + Business Media B.V., Dordrecht. https://doi.org/10.1007/978-1-4020-8524-6_1

Conti T (1999a) Quality standards development in a hypercompetitive scenario. TQM Mag 11(6):402–408. https://doi.org/10.1108/09544789910287692

Conti T (1999b) Vision 2000: positioning the new ISO 9000 standards with respect to total quality management models. Total Qual Manag 10(4–5):454–464. https://doi.org/10.1080/0954412997406

Crossland WJ (1997) HACCP and factory auditing. In: Chesworth N (ed) Food hygiene auditing. Chapman & Hall, London, pp 30–52

Daf SP, Zanwar DR (2013) Analysis and improvement in plant layout for effective production in manufacturing industries. Int J Technol 3(1):19–28

Davidson RK, Antunes W, Madslien EH, Belenguer J, Gerevini M, Torroba Perez T, Prugger R (2017) From food defence to food supply chain integrity. Brit Food J 119(1):52–66. https://doi.org/10.1108/BFJ-04-2016-0138

Deane DD, Hammond EG (1960) Coagulation of milk for cheese making by ester hydrolysis. J Dairy Sci 43(10):1421–1429. https://doi.org/10.3168/jds.s0022-0302(60)90344-1

Delgado AM, Vaz Almeida MD, Parisi S (2017) Chemistry of the mediterranean diet. Springer International Publishing, Cham

Dick GP (2000) ISO 9000 certification benefits, reality or myth? TQM Mag 12(6):365–371. https://doi.org/10.1108/09544780010351517

dos Santos Costa FD (2014) Aplicação das normas IFS, BRC e ISO 22000 na metodologia yourSTEP®. Dissertation, Universidade de Aveiro, Aveiro

Drira A, Pierreval H, Hajri-Gabouj S (2007) Facility layout problems: a survey. Ann Rev Control 31(2):255–267. https://doi.org/10.1016/j.arcontrol.2007.04.001

Ene C (2013) The relevance of traceability in the food chain. Econ Agric 60, 2:217–436. Available https://ageconsearch.umn.edu/record/152808/files/6%20-%20Ene.pdf. Accessed 10 Apr 2019

Escanciano C, Santos-Vijande ML (2014) Reasons and constraints to implementing an ISO 22000 food safety management system: evidence from Spain. Food Control 40:50–57. https://doi.org/10.1016/j.foodcont.2013.11.032

European Parliament and Council (2008) Regulation (EC) No 1334/2008 of the European Parliament and of the Council of 16 December 2008 on flavourings and certain food ingredients with flavouring properties for use in and on foods and amending Council Regulation (EEC) No 1601/91, Regulations (EC) No 2232/96 and (EC) No 110/2008 and Directive 2000/13/EC. Off J Eur Union L 354:34–50

European Parliament and Council (2011) Regulation (EU) No 1169/2011 of the European Parliament and of the Council of 25 October 2011 on the provision of food information to consumers, amending Regulations (EC) No 1924/2006 and (EC) No 1925/2006 of the European Parliament and of the Council, and repealing Commission Directive 87/250/EEC, Council Directive 90/496/EEC, Commission Directive 1999/10/EC, Directive 2000/13/EC of the European Parliament and of the Council, Commission Directives 2002/67/EC and 2008/5/EC and Commission Regulation (EC) No 608/2004. Off J Eur Union L 304(18):18–63

Fadini A, Schnepel FM (1989) Vibrational spectroscopy: methods and applications (Ellis Horwood series in analytical chemistry). Ellis Horwood Ltd, Hemel

Fennema OR, Damodaran S, Parkin KL (2017) Introduction to food chemistry. In: Damodaran S, Parkin KL (eds) Fennema's food chemistry, 5th edn. CRC Press, Boca Raton, pp 1–16

Filipović IB, Njari B, Kozačinski L, Cvrtila Fleck Ž, Mioković B, Zdolec N, Dobranić V (2008) Sustavi upravljanja kvalitetom u prehrambenoj industriji. Meso 10(6):435–438

Fitzgerald E, Buckley J (1985) Effect of total and partial substitution of sodium chloride on the quality of Cheddar cheese. J Dairy Sci 68(12):3127–3134. https://doi.org/10.3168/jds.s0022-0302(85)81217-0

Forina M, Lanteri S, Armanino C (1987) Chemometrics in food chemistry. In: Chemometrics and species identification. Top Curr Chem 141:91–143. https://doi.org/10.1007/3-540-17308-0_4

FSSC (2017) FSSC 22000 version 4.1. Foundation FSSC 22000, Gorinchem, http://www.fssc22000.com

Fulponi L (2006) Private voluntary standards in the food system: the perspective of major food retailers in OECD countries. Food Policy 31(1):1–13. https://doi.org/10.1016/j.foodpol.2005.06.006

Fulponi L, Géraud-Héraud E, Hammoudi H, Valceschini E (2006) Food safety and retailers' collective norms: impact on supply and food chains. Institut National de la Recherche Agronomique (INRA) Sciences Sociales 5–6:1–8. Available https://ageconsearch.umn.edu/record/157450/files/iss06-5-6-2_eng.pdf. Accessed 11 Apr 2019

Grover AK, Chopra S, Mosher GA (2015) Adoption of Food Safety Modernization Act: A Six Sigma Approach to Risk Based Preventive Controls for Small Food Facilities. In: Conference Proceeding of the 2015 Annual ATMAE Conference, Pittsburgh, 11–14 Nov 2015. Available https://lib.dr.iastate.edu/cgi/viewcontent.cgi?article=1464&context=abe_eng_conf. Accessed 10 Apr 2019

Grover AK, Chopra S, Mosher GA (2016) Food safety modernization act: a quality management approach to identify and prioritize factors affecting adoption of preventive controls among small food facilities. Food Control 66:241–249. https://doi.org/10.1016/j.foodcont.2016.02.001

Gurnari G (2015) The reduction of microbial spreading: little details, great effects. In: Safety protocols in the food industry and emerging concerns, pp 19–30. Springer International Publishing, Cham. https://doi.org/10.1007/978-3-319-16492-2_2

Holah J (2003) Guidelines for the hygienic design, construction and layout of food processing factories. Campden & Chorleywood, Food Research Association Group, Gloucestershire

Ibarz A, Barbosa-Cánovas GV (2002) Unit operations in food engineering. CRC Press, Boca Raton

IFS (2017) IFS food—Standard for auditing quality and food safety of food products, Version 6.1 November 2017. International Featured Standards (IFS) Management GmbH, Berlin

ISO (2018) ISO 22000:2018—Food safety management systems—Requirements for any organization in the food chain. ISO TC 34/SC 17 (Management systems for food safety). International Organization for Standardization, Geneva. Available https://www.iso.org/standard/65464.html

Johnson JD (1981) Manufacture of hard, grating cheese. US Patent Application 4,244,972, 13 Jan 1981

Johnson R (2014) Implementation of the FDA food safety modernization Act (FSMA, P.L. 111 353) Congressional Research Service Report, vol 7–5700, R43724. Congressional Research Service, Washington, pp 1–19

Kanai S, Kakizaki S, Matsutani S, Nakata H, Kaneko S (2015) Case studies of food defense in relation to plant security management. Hitachi Rev 64, 4:209–214. Available http://www.hitachi.com/rev/pdf/2015/r2015_04_105.pdf. Accessed 11 Apr 2019

Keller B, Olson NF, Richardson T (1974) Mineral retention and rheological properties of Mozzarella cheese made by direct acidification. J Dairy Sci 57(2):174–180. https://doi.org/10.3168/jds.s0022-0302(74)84856-3

Kendall H, Naughton P, Kuznesof S, Raley M, Dean M, Clark B, Stolz H, Chan MY, Zhong Q, Brereton P, Frewer LJ (2018) Food fraud and the perceived integrity of European food imports into China. PLoS ONE 13(5):e0195817. https://doi.org/10.1371/journal.pone.0195817

Knaflewska J, Pospiech E (2007) Quality assurance systems in food industry and health security of food. Acta Sci Pol Technol Aliment 6(2):75–84

Koenig J (2011) The federal food safety modernization act: impacts in import and small-scale production sectors. Harvard Library, Harvard University. Available http://nrs.harvard.edu/urn-3:HUL.InstRepos:8592051. Accessed 10 Apr 2019

Kowalska A (2018) The study of the intersection between food fraud/adulteration and authenticity. Acta Univ Agric Silvic Mendelianae Brun 66(5):1275–1286. https://doi.org/10.11118/actaun201866051275

Krieger S, Schiefer G (2007) Conception of integrated models for quality management in production chains in the agri-food sector. In: Theuvsen L, Spiller A, Peupert M, Jahn G (eds) Quality management in food chains, pp 303–313. Wageningen Academic Publishers, Wageningen. https://doi.org/10.3920/978-90-8686-605-2

Laganà P, Avventuroso E, Romano G, Gioffré ME, Patanè P, Parisi S, Moscato U, Delia S (2017) Chemistry and hygiene of food additives. Springer International Publishing, Cham. https://doi.org/10.1007/978-3-319-57042-6

Laux CM, Hurburgh CR Jr (2012). Using quality management systems for food traceability. J Ind Technol 26, 3:1–10. Available https://lib.dr.iastate.edu/cgi/viewcontent.cgi?article=1717&context=abe_eng_pubs. Accessed 11 Apr 2019

Ledford RA, O' Sullivan AC, Nath KR (1966) Residual casein fractions in ripened cheese determined by polyacrylamide-gel electrophoresis. J Dairy Sci 49, 9:1098–1101. https://doi.org/10.3168/jds.s0022-0302(66)88024-4

Lindsay RC (2007) Food additives. In: Parkin KL, Fennema OR (eds) Fennema's food chemistry. CRC Press, Boca Raton, pp 701–762

Lück E (1985) Chemical preservation of food. Zent Bakteriol Mikrobiol Hyg 1 Abt Orig B Hyg 180, 2–3:311-318

Luning PA, Devlieghere F, Verhé R (2006) Safety in the agri-food chain. Wageningen Academic Publishers, Wageningen

Lupien JR (2005) Food quality and safety: traceability and labeling. Crit Rev Food Sci Nutr 45(2):119–123. https://doi.org/10.1080/10408690490911774

Mania I, Barone C, Pellerito A, Laganà P, Parisi S (2017) Trasparenza e Valorizzazione delle Produzioni Alimentari. 'etichettatura e la Tracciabilità di Filiera come Strumenti di Tutela delle Produzioni Alimentari. Ind Aliment 56, 581:18–22

Mania I, Delgado AM, Barone C, Parisi S (2018a) Food packaging and the mandatory traceability in Europe. In: Traceability in the dairy industry in Europe. Springer International Publishing, Heidelberg, Germany. https://doi.org/10.1007/978-3-030-00446-0_8

Mania I, Delgado AM, Barone C, Parisi S (2018b) Traceability in the dairy industry in Europe. Springer International Publishing, Heidelberg, Germany. https://doi.org/10.1007/978-3-030-00446-0

Mania I, Delgado AM, Barone C, Parisi S (2018c) The ExTra tool—A practical example of extended food traceability for cheese productions. In: Traceability in the dairy industry in Europe. Springer International Publishing, Heidelberg. https://doi.org/10.1007/978-3-030-00446-0_3

Mania I, Delgado AM, Barone C, Parisi S (2018d) Food additives for analogue cheeses and traceability: the ExTra tool. In: Traceability in the dairy industry in Europe. Springer International Publishing, Heidelberg. https://doi.org/10.1007/978-3-030-00446-0_7

Martínez Fuentes C, Balbastre Benavent F, Angeles Escribá Moreno M, González Cruz T, Pardo del Val M (2000) Analysis of the implementation of ISO 9000 quality assurance systems. Work Study 49(6):229–241. https://doi.org/10.1108/00438020010343408

McEntire JC, Arens S, Bernstein M, Bugusu B, Busta FF, Cole M, Davis A, Fisher W, Geisert S, Jensen H, Kenah B, Lloyd B, Mejia C, Miller B, Mills R, Newsome R, Osho k, Prince G, Scholl S, Sutton D, Welt B, Ohlhorst S (2010) Traceability (product tracing) in food systems: an IFT report submitted to the FDA, volume 1: technical aspects and recommendations. Compr Rev Food Sci Food Saf 9, 1:92-158. https://doi.org/10.1111/j.1541-4337.2009.00097.x

McMeekin TA, Baranyi J, Bowman J, Dalgaard P, Kirk M, Ross T, Schmid S, Zwietering MH (2006) Information systems in food safety management. Int J Food Microbiol 112(3):181–194. https://doi.org/10.1016/j.ijfoodmicro.2006.04.048

Moerman F, Wouters PC (2016) Hygiene Concepts for Food Factory Design. In: Leadley CE (ed) Innovation and future trends in food manufacturing and supply chain technologies. Woodhead Publishing Ltd, Cambridge, pp 81–133

Moore JC, Spink J, Lipp M (2012) Development and application of a database of food ingredient fraud and economically motivated adulteration from 1980 to 2010. J Food Sci 77(4):R118–R126. https://doi.org/10.1111/j.1750-3841.2012.02657.x

Mukundan MK (2005) Quality and safety in Indian seafood industry. Central Institute of Fisheries Technology (CIFT) Winter School, Cochin, 14 pp. Available http://drs.cift.res.in/bitstream/handle/123456789/1244/Quality%20and%20safety%20in%20Indian%20seafood%20industry.pdf?sequence=1&isAllowed=y. Accessed 11 Apr 2019

Nikoleiski D (2015) Hygienic design and cleaning as an allergen control measure. In: Flanagan S (ed) Handbook of food allergen detection and control, Woodhead Publishing Series in Food Science, Technology and Nutrition, pp 89–102. Woodhead Publishing Ltd, Cambridge. https://doi.org/10.1533/9781782420217.1.89

Nuñez M, Medina M, Gaya P (1989) Ewes' milk cheese: technology, microbiology and chemistry. J Dairy Res 56(2):303–321. https://doi.org/10.1017/s0022029900026510

Okigbo LM, Richardson GH, Brown RJ, Ernstrom CA (1985) Effects of pH, calcium chloride, and chymosin concentration on coagulation properties of abnormal and normal milk. J Dairy Sci 68(10):2527–2533. https://doi.org/10.3168/jds.s0022-0302(85)81132-2

Opara LU, Mazaud F (2001) Food traceability from field to plate. Outlook Agric 30(4):239–247

Oser BL (1985) Highlights in the history of saccharin toxicology. Food Chem Toxicol 23(4–5):535–542. https://doi.org/10.1016/0278-6915(85)90148-6

Panisello PJ, Quantick PC (2001) Technical barriers to hazard analysis critical control point (HACCP). Food Control 12(3):165–173. https://doi.org/10.1016/S0956-7135(00)00035-9

Parisi S (2005) La produzione 'continua' è anche 'costante'? Confutazione di alcuni luoghi comuni nel settore industriale/ manifatturiero. Il Chimico Italiano XVI, 3/4, 10:18

Parisi S (2009) Intelligent packaging for the food industry. Polymer electronics—a flexible technology, Smithers Rapra Technology Ltd, Shawbury

Parisi S (2012) Food packaging and food alterations. The user-oriented approach. Smithers Rapra Technology Ltd, Shawbury

Parisi S (2016) The world of foods and beverages today: globalization, crisis management and future perspectives. Learning.ly. The Economist Group, available http://learning.ly/products/the-world-of-foods-and-beverages-today-globalization-crisis-management-and-future-perspectives. Accessed 10 Apr 2019

Parisi S (2017) Antimicrobials in foods today and the role of Chitosan—current hopes and new perspectives. Glob Drugs Therap 2(2):1–2. https://doi.org/10.15761/GDT.1000114

Parisi S (2018) Analytical approaches and safety evaluation strategies for antibiotics and antimicrobial agents in food products. Chemical and biological solutions. J AOAC Int. 101, 4:914–915. https://doi.org/10.5740/jaoacint.17-0444

Parisi S, Luo W (2018) The importance of Maillard reaction in processed foods. Springer International Publishing, Cham. https://doi.org/10.1007/978-3-319-95463-9

Quinn MJ Jr, Ziolkowski D Jr (2015) Wildlife toxicity assessment for triacetin. In: Williams MA, Reddy G, Quinn MJ Jr, Johnson MS, Wildlife toxicity assessments for chemicals of military concern, p. 291–301. Elsevier, Amsterdam, Waltham, and Oxford. https://doi.org/10.1016/b978-0-12-800020-5.00017-x

Renzi MF, Cappelli L (2000) Integration between ISO 9000 and ISO 14000: opportunities and limits. Total Qual Manag 11, 4–6:849-856. doi: https://doi.org/10.1080/09544120050008318

RSPO (2014) RSPO supply chain certification standard. The Roundtable on Sustainable Palm Oil (RSPO), RSPO Malaysia, Kuala Lumpur, and RSPO Indonesia, Jakarta Selatan. Available https://www.rspo.org/articles/download/217a67fe502c19d. Accessed 10 Apr 2019

Sagara K, Ueda K, Shimada T, Ishii T (1990) Continuous production process of cheese curds and production process of cheese therefrom. US Patent Application 4,948,599, 14 Aug 1990

Saltmarsh M (ed) (2000) Essential guide to food additives. Leatherhead Publishing, LFRA Ltd., Leatherhead

Schulze H, Albersmeier F, Gawron JC, Spiller A, Theuvsen L (2008) Heterogeneity in the evaluation of quality assurance systems: the International Food Standard (IFS) in European agribusiness. Int Food Agribus Manag Rev 119(1):52–66. https://doi.org/10.1108/BFJ-04-2016-0138

Shahin A, Poormostafa M (2011) Facility layout simulation and optimization: an integration of advanced quality and decision making tools and techniques. Mod Appl Sci 5(4):95–111. https://doi.org/10.5539/mas.v5n4p95

Shaw M (1984) Subject: Novel cheeses and cheese-making processes. Int J Dairy Technol 37(1):27–31. https://doi.org/10.1111/j.1471-0307.1984.tb02278.x

Shehata AE, Iyer M, Olson NF, Richardson T (1967) Effect of type of acid used in direct acidification procedures on moisture, firmness, and calcium levels of cheese. J Dairy Sci 50(6):824–827. https://doi.org/10.3168/jds.s0022-0302(67)87529-5

Singhal RS, Kulkarni PK, Reg DV (eds) (1997) Handbook of indices of food quality and authenticity. Woodhead Publishing Ltd, Cambridge

Singla RK, Dubey AK, Ameen SM, Montalto S, Parisi S (2018) The control of Maillard reaction in processed foods. Analytical testing methods for the determination of 5-Hydroxymethylfurfural. In: Analytical methods for the assessment of Maillard reactions in foods. Springer briefs in molecular science. Springer, Cham. https://doi.org/10.1007/978-3-319-76923-3_2

Sinha S (2014) Creating a culture of compliance across your supply chain. Qual Dig. https://www.qualitydigest.com. Available https://www.qualitydigest.com/inside/quality-insider-article/creating-culture-compliance-across-your-supply-chain.html. Accessed 11 Apr 2019

SQF (2017) SQF quality code, edition 8. Safe quality food (SQF) Institute, Arlington. Available https://www.sqfi.com/what-is-the-sqf-program/sqf-quality-program/. Accessed 10 Apr 2019

SSICA, Certiquality, Mobilpesca Surgelati SpA (2009) Linea guida ISO 22000 SGSA di prodotti ittici. Stazione Sperimentale per l'Industria delle Conserve Alimentari (SSICA), Parma, Certiquality Srl, Milan, and Mobilpesca Surgelati SpA, Altopascio. Available http://www.ssica.it/component/option,com_docman/task,doc_download/gid,271/lang,it/. Accessed 11 Apr 2019

Stein K (2015) Effective allergen management practices to reduce allergens in food. In Handbook of food allergen detection and control. In: Flanagan S (ed) Handbook of food allergen detection and control, Woodhead Publishing Series in Food Science, Technology and Nutrition, pp 103–131. Woodhead Publishing Ltd, Cambridge. https://doi.org/10.1533/9781782420217.1.103

Steinka I, Barone C, Parisi S, Micali M (2017) Technology and chemical features of frozen vegetables. In: The chemistry of frozen vegetables, pp 23–29. Springer Briefs in Molecular Science. Springer, Cham. https://doi.org/10.1007/978-3-319-53932-4_2

Stilo A, Parisi S, Delia S, Anastasi F, Bruno G, Laganà P (2009) Food security in Europe: comparison between the "Hygiene Package" and the British Retail Consortium (BRC) & International Food Standard (IFS) protocols. Ann Ig 21(4):387–401

Sun YM, Ockerman HW (2005) A review of the needs and current applications of hazard analysis and critical control point (HACCP) system in foodservice areas. Food Control 16(4):325–332. https://doi.org/10.1016/j.foodcont.2004.03.012

Szpylka J, Thiex N, Acevedo B, Albizu A, Angrish P, Austin S, Bach Knudsen KE, Barber CA, Berg D, Bhandari SD, Bienvenue A, Cahill K, Caldwell J, Campargue C, Cho F, Collison MW, Cornaggia C, Cruijsen H, Das M, De Vreeze M, Deutz I, Donelson J, Dubois A, Duchateau GS, Duchateau L, Ellingson D, Gandhi J, Gottsleben F, Hache J, Hagood G, Hamad M, Haselberger PA, Hektor T, Hoefling R, Holroyd S, Holt DL, Horst JG, Ivory R, Jaureguibeitia A, Jennens M, Kavolis DC, Kock L, Konings EJM, Krepich S, Krueger DA, Lacorn M, Lassitter CL, Lee S, Li H, Liu A, Liu K, Lusiak BD, Lynch E, Mastovska K, McCleary BV, Mercier GM, Metra PL, Monti L, Moscoso CJ, Narayanan H, Parisi S, Perinello G, Phillips MM, Pyatt S, Raessler M, Reimann LM, Rimmer CA, Rodriguez A, Romano J, Salleres S, Sliwinski M, Smyth G, Stanley K, Steegmans M, Suzuki H, Swartout K, Tahiri N, Ten Eyck R, Torres Rodriguez MG, Van Slate J, Van Soest PJ, Vennard T, Vidal R, Hedegaard RSV, Vrasidas I, Vrasidas Y, Walford S, Wehling P, Winkler P, Winter R, Wirthwine B, Wolfe D, Wood L, Woollard DC, Yadlapalli S, Yan X, Yang J, Yang Z, Zhao G. (2018a) Standard Method Performance Requirements (SMPRs®) 2018.001: Sugars in Animal Feed, Pet Food, and Human Food. J AOAC Int101, 4:1280–1282. https://doi.org/10.5740/jaoacint.smpr2018.001

Szpylka J, Thiex N, Acevedo B, Albizu A, Angrish P, Austin S, Bach Knudsen KE, Barber CA, Berg D, Bhandari SD, Bienvenue A, Cahill K, Caldwell J, Campargue C, Cho F, Collison MW, Cornaggia C, Cruijsen H, Das M, De Vreeze M, Deutz I, Donelson J, Dubois A, Duchateau GS, Duchateau L, Ellingson D, Gandhi J, Gottsleben F, Hache J, Hagood G, Hamad M, Haselberger PA, Hektor T, Hoefling R, Holroyd S, Lloyd Holt D, Horst JG, Ivory R, Jaureguibeitia A, Jennens M, Kavolis DC, Kock L, Konings EJM, Krepich S, Krueger DA, Lacorn M, Lassitter CL, Lee S, Li H, Liu A, Liu K, Lusiak BD, Lynch E, Mastovska K, McCleary BV, Mercier GM, Metra PL, Monti L, Moscoso CJ, Narayanan H, Parisi S, Perinello G, Phillips MM, Pyatt S, Raessler M, Reimann LM, Rimmer CA, Rodriguez A, Romano J, Salleres S, Sliwinski M, Smyth G, Stanley K, Steegmans M, Suzuki H, Swartout K, Tahiri N, Eyck RT, Torres Rodriguez MG, Van Slate J, Van Soest PJ, Vennard T, Vidal R, Vinbord Hedegaard RS, Vrasidas I, Vrasidas Y, Walford S, Wehling P, Winkler P, Winter R, Wirthwine B, Wolfe D, Wood L, Woollard DC, Yadlapalli S, Yan X, Yang J, Yang Z, Zhao G. (2018b) Standard Method Performance Requirements (SMPRs®) 2018.002: Fructans in Animal Food (Animal Feed, Pet Food, and Ingredients). J AOAC Int101, 4:1283–1284. https://doi.org/10.5740/jaoacint.smpr2018.002

Tanner B (2000) Independent assessment by third-party certification bodies. Food Control 11(5):415–417. https://doi.org/10.1016/S0956-7135(99)00055-9

Tian F (2017) A supply chain traceability system for food safety based on HACCP, blockchain & Internet of things. In: Proceedings of the 2017 international conference on service systems and service management, Dalian, China, 16–18 June 2017 pp 1–6. https://doi.org/10.1109/icsssm.2017.7996119

Torres N, Chandan RC (1981) Latin American white cheese—a review. J Dairy Sci 64(3):552–557. https://doi.org/10.3168/jds.s0022-0302(81)82608-2

Trecker GW, Monckton SP (1990) Grated hard Parmesan cheese and method for making same. US Patent Application 4,960,605, 02 Nov 1990

Trienekens J, Zuurbier P (2008) Quality and safety standards in the food industry, developments and challenges. Int J Prod Econ 113(1):107–122. https://doi.org/10.1016/j.ijpe.2007.02.050

Tsimidou M, Macrae R, Wilson I (1987) Authentication of virgin olive oils using principal component analysis of triglyceride and fatty acid profiles: Part 2—Detection of adulteration with other vegetable oils. Food Chem 25(4):251–258. https://doi.org/10.1016/0308-8146(87)90011-2

UNECE (1981) Development of airborne equipment to intensify world food production. United Nations Economic Commission for Europe (UNECE), Geneva

Van der Spiegel M, Luning PA, Ziggers GW, Jongen WMF (2003) Towards a conceptual model to measure effectiveness of food quality systems. Trends Food Sci Technol 14(10):424–431. https://doi.org/10.1016/S0924-2244(03)00058-X

Van Donk DP, Gaalman G (2004) Food Safety and Hygiene. Chem Eng Res Des 82(11):1485–1493. https://doi.org/10.1205/cerd.82.11.1485.52037

Vloeberghs D, Bellens J (1996) Implementing the ISO 9000 standards in Belgium. Qual Prog 29(6):43–48

Waldron KW (ed) (2009) Handbook of waste management and co-product recovery in food processing, vol 2. Woodhead Publishing Ltd., Oxford, Cambridge, and New Delhi

Wanniarachchi WNC, Gopura RARC, Punchihewa HKG (2016) Development of a layout model suitable for the food processing industry. J Ind Eng Article ID 2796806. https://doi.org/10.1155/2016/2796806

Wargel RJ, Greiner SP, Hettinga DH (1981) Process and products for the manufacture of cheese flavored products. US Patent Application 4,244,971, 13 Jan 1981

Williams AA (1985) Methods used in the authentication of wines. In: Birch GG, Lindley MG (eds) Alcoholic beverages. Chapman and Hall, London

Wognum PN, Bremmers H, Trienekens JH, van der Vorst JG, Bloemhof JM (2011) Systems for sustainability and transparency of food supply chains—Current status and challenges. Adv Eng Inform 25(1):65–76. https://doi.org/10.1016/j.aei.2010.06.001

Woodroof JG (1966) Peanuts: production, processing products. Peanuts: production, processing products. Avi Publishing Co., Westport

Wu L, Xu L, Gao J (2011) The acceptability of certified traceable food among Chinese consumers. Brit Food J 113(4):519–534. https://doi.org/10.1108/00070701111123998

Yung WK (1997) The values of TQM in the revised ISO 9000 quality system. Int J Op Prod Manag 17(2):221–230. https://doi.org/10.1108/01443579710158078

Zaccheo A, Palmaccio E, Venable M, Locarnini-Sciaroni I, Parisi S (2017) Food hygiene and applied food microbiology in an anthropological cross cultural perspective. Springer International Publishing, Cham

Zarling P (2018) Consumers crave authentic ethnic food—and will pay more. Food dive, Washington. Available https://www.fooddive.com/news/consumers-crave-authentic-ethnic-food-and-will-pay-more/529282/. Accessed 08 Apr 2019

Chapter 2
The Intentional Adulteration in Foods and Quality Management Systems: Chemical Aspects

Abstract The problem food frauds today is correlated with the identification of main and minor ingredients for food and beverage productions. Adulteration is a big concern in food environments, and a strong answer is needed by the whole world of food and beverage stakeholders. The existence of quality certification standards makes possible the management of food frauds and other menaces concerning 'identity loss' with relation to four main factors: the role of food additives, processing aids and chemical compounds in general for food and beverage production; adequate actions and mitigation strategies against food frauds, both intentional and unavoidable events; the assessment of documentation concerning food productions by external and experienced auditors; and possible safety risks caused by the consumption of undeclared allergens. With concern to food quality management systems, mitigation plans against food frauds are one of the most challenging areas. The aim of this chapter is to give a detailed idea of basic requirements in this ambit by the viewpoint of food quality management schemes.

Keywords Allergen · Audit · GSFS · IFS food · Intentional adulteration · Quality system · Traceability

Abbreviations

BRC	British Retail Consortium
EMA	Economically motivated adulteration
EU	European Union
FDA	Food and Drug Administration
FBO	Food Business Operator
FSSC	Food Safety System Certification
GFSI	Global Food Safety Initiative
GSFS	Global Standard for Food Safety
HACCP	Hazard Analysis and Critical Control Points
HARPC	Hazard Analysis and Risk-based Preventive Controls
IFS	International Featured Standard

© The Author(s), under exclusive license to Springer Nature Switzerland AG 2019
M. Fiorino et al., *Quality Systems in the Food Industry*, Chemistry of Foods,
https://doi.org/10.1007/978-3-030-22553-7_2

ISO   International Organization for Standardization
RASFF   Rapid Alert System for Food and Feed
RSPO   Roundtable on Sustainable Palm Oil
SQF   Safe Quality Food

2.1 Intentional Adulteration in Food Industries and Quality Systems

By a general viewpoint, the problem food frauds today is correlated with the identi-fication of main and minor ingredients for food and beverage productions. In detail, it should be recognised that 'adulteration' is a big concern in food environments (Manning 2016; Manning and Soon 2014, 2016; Silvis et al. 2017; Spink et al. 2016; Van Ruth et al. 2017), and a strong answer is needed by the whole world of food and beverage stakeholders (not only food and business operators, but also official institutions, training and research institutes and so on).[1]

One of the most interesting countermeasures taken into account so far is offered by quality certification organisations. The existence of quality certification standards makes now possible the management of food frauds and other menaces concerning 'identity loss' with relation to four main factors:

(1) Food additives, processing aids and chemical compounds in general for food and beverage production are involved when speaking of traceability, authenticity and access to the production in flow diagrams (Asensio et al. 2008; Barbieri et al. 2014; Delgado et al. 2017; Mania et al. 2018a, b, c; Lindsay 2007; Lupien 2005; McEntire et al. 2010; Singhal et al. 1997)

(2) Adequate actions against food frauds, both intentional and unavoidable events, need a preliminary risk assessment, determined mitigation strategies, a powerful supplier evaluation (including related monitoring actions), new or ameliorated chemical and biological analyses and in-depth analysis of markets and economic advantages/risk balances when speaking of intentional adulteration (Fennema et al. 2017; Moore et al. 2012; Johnson 2014)

(3) The assessment of documentation concerning food productions by external and experienced auditors is a critical point (Crossland 1997; Panisello and Quantick 2001)

(4) Finally, adulteration is not always synonym of food safety risk. On the other hand, the loss of identity can surely enhance possible health risks such as aller-genic and unpredicted reactions (BRC 2017; Nikoleiski 2015; Stein 2015).

As a premise, countermeasures against food frauds (Sect. 1.2) concern many areas of the so-called quality management areas had identified in the ambit of food quality

---

[1] These and other information are discussed in the 'Food Traceability' lecture for the Faculty of Agricultural Technology, Al-Balqa Applied University, Al-Salt, Jordan, by Salvatore Parisi, first course 2018–2019.

standards. The most known of these standards are at present (BRC 2018a; FSSC 2017; IFS 2017; ISO 2018; RSPO 2014; SQF 2017):

- The Global Standard for Food Safety (GSFS) by the British Retail Consortium (BRC)
- The International Featured Standard (IFS) Food, by the IFS
- The ISO 22000 norm, by the ISO
- The Food Safety System Certification (FSSC) 22000
- The Safe Quality Food (SQF) by the SQF Institute, USA
- The Roundtable on Sustainable Palm Oil (RSPO) certification scheme, by the homonym organisation.

It should be highlighted that these standards, based on the ISO 9000 series of standards by the International Organization for Standardization (ISO), in particular ISO 9001:2015 (ISO 9015), have been elaborated with the aim of giving a strong adherence to the food and beverage sector as really different from other non-food ambits. In general, the goal of ISO was to give all possible companies and institutions—including food and beverage companies also—some basic and reliable instrument for achieving the following results: the reliable design, creation, delivery and continuous improvement of supplied products and services (Brown et al. 1998; Buttle 1997; Dick 2000; Martínez Fuentes et al. 2000; Vloeberghs and Bellens 1996; Yung 1997).

However, food and beverage industries are different from non-food environments such as the production of oil derivatives or plastic matters because of different risks and correlated counterstrategies (Baines 1999; Channaiah et al. 2017; Davidson et al. 2017; Grover et al. 2015; Parisi 2009, 2012, 2016; Gurnari 2015; Schulze et al. 2008; Stilo et al. 2009), according to the Hazard Analysis and Critical Control Points (HACCP) assessment, and the Hazard Analysis and Risk-based Preventive Controls (HARPC) (Baines 1999; Channaiah et al. 2017; Davidson et al. 2017; Grover et al. 2015; Parisi 2009, 2012, 2016; Gurnari 2015; Schulze et al. 2008; Stilo et al. 2009).

With concern to above-mentioned quality systems, the following areas of intervention and improvement are generally considered when speaking of contrast to food frauds (BRC 2018b; Parisi and Luo 2018; Singla et al. 2018; Steinka et al. 2017; Zaccheo et al. 2017):

(1) The commitment of senior management
(2) The research for continuous improvement
(3) The 'food safety plan', including aspects concerning processing operations and related flow charts, site design, segregation and sanification procedures; training and attention to safety culture
(4) Practical actions of the management in the organisation: internal audits, non-conformities and preventive/corrective actions
(5) The assessment and re-evaluation of all raw material suppliers
(6) The creation, implementation, improvement and continuous demonstrability of traceability and labelling requirements

However, the basic question in this ambit should concern intentional or unavoidable food frauds, because of the possible occurrence of both events when speaking of a single situation (naturally, on the legal ground). Consequently, a reliable definition of both food fraud events should be discussed.

2.2 What Is Intentional Adulteration? Some Clarifications by the Quality Management Viewpoint

According to the GSFS Issue 8 (BRC 2018b), adulteration is considered a synonym word for 'fraud'. In detail, food frauds should be considered (clause 2.7.1) as substitution of one raw material, food additive, processing aid or intentional adulteration of a specific food or beverage product. According to this interpretation, the HACCP or HARPC study has to consider different risks for assessment, including the potential menace for fraud and/or adulteration, meaning that 'fraud' and 'adulteration' are not the same thing.

The definition of 'fraud' can be stated in different ways: according to the GSFS Issue 8, 'food fraud' corresponds to the intentional and undeclared substitution, dilution or addition of a raw material (including additives, processing aids and so on) to the product. A different form of fraud is represented when speaking of intentional misrepresentation of the final product or the used ingredient(s) with demonstrable economic advantages (BRC 2018b). The IFS Food standard, version 6.1, states that food fraud concerns not only edible ingredients, but also 'packaging placed upon the market for economic gain'. In addition, the definition of food fraud has to cover also outsourced processes (IFS 2017).

In the GSFS ambit, the definition of 'intentional adulteration' or 'economically motivated adulteration' (EMA) should be 'the addition of an undeclared material into a food item or raw material for economic gain'. In addition, 'adulterant' is defined as 'an undeclared material added into a food item or raw material for economic gain'. The Food and Drug Administration (FDA) definition for EMA in 2009 clearly stated that EMA should be considered as the 'fraudulent, intentional substitution or addition of a substance in a product for the purpose of increasing the apparent value of the product or reducing the cost of its production, i.e., for economic gain' (Johnson 2014).

With reference to GSFS and IFS Food interpretations, the safety plan has to consider new or emerging issues when speaking of 'known adulteration', such as news from the Rapid Alert System for Food and Feed (RASFF) Portal of the European Union (EU) (Parisi et al. 2016).[2] Actually, these reports may be also normal news from newspapers, information agencies, trade associations, etc.: the important thing is that a possible adulteration risk is known and that adequate countermeasures are taken into account as soon as possible (clause 5.4.1). The preventive approach to

---

[2]The RSSFF Portal is freely accessible at the following Internet address: https://ec.europa.eu/food/safety/rasff_en.

food safety requires (clause 5.4.1 of the GSFS) that a database of historical studies and correlated menaces is always available with the aim of minimising all possible risks caused by entering fraudulent materials in the supply chain (BRC 2018b).

Consequently, the risk assessment with relation to authenticity issues (also named 'vulnerability assessment, GSFS clause 5.4.2 and IFS Food clauses 4.21.1 and 5.6.8) has to be always available, be reviewed periodically, and take into account not only demonstrated data related to substitution and adulteration of raw materials, but also (IFS 2017)[3]:

(1) Economic studies related to advantages linked with frauds (e.g. ration between costs and assumed value)
(2) Geographical origin of the raw material (including also considerations concerning the economic region where the country is located)
(3) Features of the supply chain (number of stakeholders, number of 'nodes' and 'hubs', length of the networks, absence or presence of centralised 'traceability managers', etc.)
(4) Easy entering opportunities for fraudulent (and non-fraudulent) ingredients in the supply chain
(5) Analytical results
(6) And—naturally—the chemical and physical nature of all raw materials.

It has also be noted that the preventive approach according to GSFS requires, in the broad ambit of internal audits, the elaboration and review of a food fraud prevention plan and also a food defence plan at the same time (clause 3.4.1). This simple request means clearly that all possible food fraud episodes and related menaces can be associated with malicious attack against the food business operator (FBO). In other terms, the intentional adulteration can be 'intentional' for one or more specific players, while the producing and unaware FBO may be also seriously damaged (no advantages)! On the other hand, IFS Food version 6.1 clearly defines 'food defence' as 'the protection of food products from intentional contamination or adulteration by biological, chemical, physical or radiological agents for the purpose of causing harm', stating clearly that food defence is explicitly created and implemented against food frauds (IFS 2017). Also, the IFS Food definition of defence assessment (clause 6.1.2) states that this instrument is designed against all factors affecting food integrity, including clearly all possible EMA episodes (IFS 2017).

Finally, it should be noted that the vulnerability assessment should give some evaluation—in terms of attractiveness for intentional adulteration—for each ingredient (grades can range from very high risk of adulteration to negligible risks) (BRC 2018b). On these bases, and with relation to adequate supplier surveillance considering also frauds and intentional/unintentional adulteration episodes (GSFS clause 9.1.1; IFS Food clause 4.21.1), the company can really and reliably assure the so-called product identity or complete correspondence between the claimed and marketed food and the real sold product (BRC 2018b). For clarity purposes, the Global

---

[3]These and other information are discussed in the 'Food Traceability' lecture for the Faculty of Agricultural Technology, Al-Balqa Applied University, Al-Salt, Jordan, by Salvatore Parisi, first course 2018–2019.

Food Safety Initiative (GFSI) considers 'product identity' with relation to 'products that are of the nature, substance and quality expected' with relation to the following and undeclared alterations: substitution, dilution, adulteration and misrepresentation (while IFS prefers 'counterfeiting') (IFS 2017). The final step of this process is the so-called mitigation plan (IFS Food version 6.1): this plan is the instrument put in place with the aim of defining 'measures and controls that are required to be in place to effectively mitigate the identified risks' when speaking of food frauds. It should be based on above-mentioned elements, and require different analytical methods for preventing EMA episodes, belonging to the technical (chemical, microbiological and sensorial) area and to the economic and political ambit of the application when speaking of marketed items (Szpylka et al. 2018a, b).

References

Asensio L, González I, García T, Martín R (2008) Determination of food authenticity by enzyme-linked immunosorbent assay (ELISA). Food Control 19(1):1–8. https://doi.org/10.1016/j.foodcont.2007.02.010

Baines RN (1999) Environmental & animal safety dimensions to developing food safety & quality assurance initiatives in the United Kingdom. Market rights and equity: food and agriculture standards in a shrinking world. Institute for Food & Agricultural Standards, Michigan

Barbieri G, Barone C, Bhagat A, Caruso G, Conley ZR, Parisi S (2014) The influence of chemistry on new foods and traditional products. Springer briefs in chemistry of foods. Springer International Publishing, Heidelberg, Germany

BRC (2017) Global standard food safety, issue 7—An introduction to best-practice lubrication procedures in the food industry. British Retail Consortium (BRC) Global Standards, London, www.brcglobalstandards.com. Available https://www.brcgs.com/media/165616/brc-lubrication-guide-screen.pdf. Accessed 10 Apr 2019

BRC (2018a) Global standard food safety, issue 8. British Retail Consortium (BRC) Global Standards, London, www.brcglobalstandards.com

BRC (2018b) Global standard food safety, issue 8. Interpretation Guideline. British Retail Consortium (BRC) Global Standards, London, www.brcglobalstandards.com GSFS interpret

Brown A, Van Der Wiele T, Loughton K (1998) Smaller enterprises' experiences with ISO 9000. Int J Qual Reliab Manag 15(3):273–285. https://doi.org/10.1108/02656719810198935

Buttle F (1997) ISO 9000: marketing motivations and benefits. Int J Qual Reliab Manag 14(9):936–947. https://doi.org/10.1108/02656719710186867

Channaiah LH, Michael M, Acuff JC, Phebus RK, Thippareddi H, Olewnik M, Milliken G (2017) Validation of the baking process as a kill-step for controlling Salmonella in muffins. Int J Food Microbiol 250:1–6. https://doi.org/10.1016/j.ijfoodmicro.2017.03.007

Crossland WJ (1997) HACCP and factory auditing. In: Chesworth N (ed) Food hygiene auditing. Chapman & Hall, London, pp 30–52

Davidson RK, Antunes W, Madslien EH, Belenguer J, Gerevini M, Torroba Perez T, Prugger R (2017) From food defence to food supply chain integrity. Brit Food J 119(1):52–66. https://doi.org/10.1108/BFJ-04-2016-0138

Delgado AM, Vaz Almeida MD, Parisi S (2017) Chemistry of the mediterranean diet. Springer International Publishing, Cham

Dick GP (2000) ISO 9000 certification benefits, reality or myth? TQM Mag 12(6):365–371. https://doi.org/10.1108/09544780010351517

Fennema OR, Damodaran S, Parkin KL (2017) Introduction to food chemistry. In: Damodaran S, Parkin KL (eds) Fennema's food chemistry, 5th edn. CRC Press, Boca Raton, pp 1–16

FSSC (2017) FSSC 22000 version 4.1. Foundation FSSC 22000, Gorinchem. http://www.fssc22000.com

Grover AK, Chopra S, Mosher GA (2015) Adoption of food safety modernization act: a six sigma approach to risk based preventive controls for small food facilities. In: Conference proceeding of the 2015 annual ATMAE conference, Pittsburgh, 11–14 Nov 2015. Available https://lib.dr.iastate.edu/cgi/viewcontent.cgi?article=1464&context=abe_eng_conf. Accessed 10 Apr 2019

Gurnari G (2015) The reduction of microbial spreading: little details, great effects. In: Safety protocols in the food industry and emerging concerns, pp 19–30. Springer International Publishing, Cham. https://doi.org/10.1007/978-3-319-16492-2_2

IFS (2017) IFS food—Standard for auditing quality and food safety of food products, version 6.1, Nov 2017. International Featured Standards (IFS) Management GmbH, Berlin

ISO (2018) ISO 22000:2018—Food safety management systems—Requirements for any organization in the food chain. ISO TC 34/SC 17 (Management systems for food safety). International Organization for Standardization, Geneva. Available https://www.iso.org/standard/65464.html

Johnson R (2014) Implementation of the FDA food safety modernization Act (FSMA, P.L. 111 353) congressional research service report, vol 7–5700, R43724. Congressional Research Service, Washington, pp 1–19

Lindsay RC (2007) Food additives. In: Parkin KL, Fennema OR (eds) Fennema's food chemistry. CRC Press, Boca Raton, pp 701–762

Lupien JR (2005) Food quality and safety: traceability and labeling. Crit Rev Food Sci Nutr 45(2):119–123. https://doi.org/10.1080/10408690490911774

Mania I, Delgado AM, Barone C, Parisi S (2018a) Food packaging and the mandatory traceability in Europe. In: Traceability in the dairy industry in Europe. Springer International Publishing, Heidelberg, Germany. https://doi.org/10.1007/978-3-030-00446-0_8

Mania I, Delgado AM, Barone C, Parisi S (2018b) The ExTra tool—A practical example of extended food traceability for cheese productions. In: Traceability in the dairy industry in Europe. Springer International Publishing, Heidelberg. https://doi.org/10.1007/978-3-030-00446-0_3

Mania I, Delgado AM, Barone C, Parisi S (2018c) Food additives for analogue cheeses and traceability: The ExTra tool. In: Traceability in the dairy industry in Europe. Springer International Publishing, Heidelberg, Germany. https://doi.org/10.1007/978-3-030-00446-0_7

Manning L (2016) Food fraud: policy and food chain. Curr Opin Food Sci 10:16–21. https://doi.org/10.1016/j.cofs.2016.07.001

Manning L, Soon JM (2014) Developing systems to control food adulteration. Food Policy 49(1):23–32. https://doi.org/10.1016/j.foodpol.2014.06.005

Manning L, Soon JM (2016) Food safety, food fraud, and food defense: a fast evolving literature. J Food Sci 81(4):R823–R834. https://doi.org/10.1111/1750-3841.13256

Martínez Fuentes C, Balbastre Benavent F, Angeles Escribá Moreno M, González Cruz T, Pardo del Val M (2000) Analysis of the implementation of ISO 9000 quality assurance systems. Work Study 49(6):229–241. https://doi.org/10.1108/00438020010343408

McEntire JC, Arens S, Bernstein M, Bugusu B, Busta FF, Cole M, Davis A, Fisher W, Geisert S, Jensen H, Kenah B, Lloyd B, Mejia C, Miller B, Mills R, Newsome R, Osho k, Prince G, Scholl S, Sutton D, Welt B, Ohlhorst S (2010) Traceability (product tracing) in food systems: an IFT report submitted to the FDA, volume 1: technical aspects and recommendations. Compr Rev Food Sci Food Saf 9, 1:92–158. https://doi.org/10.1111/j.1541-4337.2009.00097.x

Moore JC, Spink J, Lipp M (2012) Development and application of a database of food ingredient fraud and economically motivated adulteration from 1980 to 2010. J Food Sci 77(4):R118–R126. https://doi.org/10.1111/j.1750-3841.2012.02657.x

Nikoleiski D (2015) Hygienic design and cleaning as an allergen control measure. In: Flanagan S (ed) Handbook of food allergen detection and control, Woodhead Publishing series in food science, technology and nutrition, pp 89–102. Woodhead Publishing Ltd, Cambridge. https://doi.org/10.1533/9781782420217.1.89

Panisello PJ, Quantick PC (2001) Technical barriers to hazard analysis critical control point (HACCP). Food Control 12(3):165–173. https://doi.org/10.1016/S0956-7135(00)00035-9

Parisi S (2009) Intelligent packaging for the food industry. Polymer electronics—a flexible technology. Smithers Rapra Technology Ltd, Shawbury

Parisi S (2012) Food packaging and food alterations. The user-oriented approach. Smithers Rapra Technology Ltd, Shawbury

Parisi S (2016) The world of foods and beverages today: globalization, crisis management and future perspectives. Learning.ly. The Economist Group, available http://learning.ly/products/the-world-of-foods-and-beverages-today-globalization-crisis-management-and-future-perspectives. Accessed 10 Apr 2019

Parisi S, Barone C, Sharma R (2016). Chemistry and food safety in the EU. Springer International Publishing, Cham. https://doi.org/10.1007/978-3-319-33393-9

Parisi S, Luo W (2018) The importance of Maillard reaction in processed foods. Springer International Publishing, Cham. https://doi.org/10.1007/978-3-319-95463-9

RSPO (2014) RSPO supply chain certification standard. The roundtable on sustainable palm oil (RSPO), RSPO Malaysia, Kuala Lumpur, and RSPO Indonesia, Jakarta Selatan. Available https://www.rspo.org/articles/download/217a67fe502c19d. Accessed 10 Apr 2019

Schulze H, Albersmeier F, Gawron JC, Spiller A, Theuvsen L (2008) Heterogeneity in the evaluation of quality assurance systems: the International Food Standard (IFS) in European agribusiness. Int Food Agribus Manag Rev 119(1):52–66. https://doi.org/10.1108/BFJ-04-2016-0138

Silvis ICJ, Van Ruth SM, Van Der Fels-klerx HJ, Luning PA (2017) Assessment of food fraud vulnerability in the spices chain: An explorative study. Food Control 81:80–87. https://doi.org/10.1016/j.foodcont.2017.05.019

Singhal RS, Kulkarni PK, Reg DV (eds) (1997) Handbook of indices of food quality and authenticity. Woodhead Publishing Ltd, Cambridge

Singla RK, Dubey AK, Ameen SM, Montalto S, Parisi S (2018) The control of Maillard reaction in processed foods. Analytical testing methods for the determination of 5-Hydroxymethylfurfural. In: Analytical methods for the assessment of Maillard reactions in foods. Springer Briefs in molecular science. Springer, Cham. https://doi.org/10.1007/978-3-319-76923-3_2

Spink J, Moyer DC, Speier-Pero C (2016) Introducing the food fraud initial screening model (FFIS). Food Control 69:306–314. https://doi.org/10.1016/j.foodcont.2016.03.016

SQF (2017) SQF quality code, edition 8. Safe Quality Food (SQF) Institute, Arlington. Available https://www.sqfi.com/what-is-the-sqf-program/sqf-quality-program/. Accessed 10 Apr 2019

Stein K (2015) Effective allergen management practices to reduce allergens in food. In: Flanagan S (ed) Handbook of food allergen detection and control, Woodhead Publishing series in food science, technology and nutrition, pp 103–131. Woodhead Publishing Ltd, Cambridge. https://doi.org/10.1533/9781782420217.1.103

Steinka I, Barone C, Parisi S, Micali M (2017) Technology and chemical features of frozen vegetables. In: The chemistry of frozen vegetables. Springer briefs in molecular science, pp 23–29. Springer, Cham. https://doi.org/10.1007/978-3-319-53932-4_2

Stilo A, Parisi S, Delia S, Anastasi F, Bruno G, Laganà P (2009) Food security in Europe: comparison between the "Hygiene Package" and the British Retail Consortium (BRC) & International Food Standard (IFS) protocols. Ann Ig 21(4):387–401

Szpylka J, Thiex N, Acevedo B, Albizu A, Angrish P, Austin S, Bach Knudsen KE, Barber CA, Berg D, Bhandari SD, Bienvenue A, Cahill K, Caldwell J, Campargue C, Cho F, Collison MW, Cornaggia C, Cruijsen H, Das M, De Vreeze M, Deutz I, Donelson J, Dubois A, Duchateau GS, Duchateau L, Ellingson D, Gandhi J, Gottsleben F, Hache J, Hagood G, Hamad M, Haselberger PA, Hektor T, Hoefling R, Holroyd S, Holt DL, Horst JG, Ivory R, Jaureguibeitia A, Jennens M, Kavolis DC, Kock L, Konings EJM, Krepich S, Krueger DA, Lacorn M, Lassitter CL, Lee S, Li H, Liu A, Liu K, Lusiak BD, Lynch E, Mastovska K, McCleary BV, Mercier GM, Metra PL, Monti L, Moscoso CJ, Narayanan H, Parisi S, Perinello G, Phillips MM, Pyatt S, Raessler M, Reimann LM, Rimmer CA, Rodriguez A, Romano J, Salleres S, Sliwinski M, Smyth G, Stanley K, Steegmans M, Suzuki H, Swartout K, Tahiri N, Ten Eyck R, Torres Rodriguez MG, Van Slate

J, Van Soest PJ, Vennard T, Vidal R, Hedegaard RSV, Vrasidas I, Vrasidas Y, Walford S, Wehling P, Winkler P, Winter R, Wirthwine B, Wolfe D, Wood L, Woollard DC, Yadlapalli S, Yan X, Yang J, Yang Z, Zhao G. (2018a) Standard method performance requirements (SMPRs®) 2018.001: sugars in animal feed, pet food, and human food. J AOAC Int101, 4:1280–1282. https://doi.org/10.5740/jaoacint.smpr2018.001

Szpylka J, Thiex N, Acevedo B, Albizu A, Angrish P, Austin S, Bach Knudsen KE, Barber CA, Berg D, Bhandari SD, Bienvenue A, Cahill K, Caldwell J, Campargue C, Cho F, Collison MW, Cornaggia C, Cruijsen H, Das M, De Vreeze M, Deutz I, Donelson J, Dubois A, Duchateau GS, Duchateau L, Ellingson D, Gandhi J, Gottsleben F, Hache J, Hagood G, Hamad M, Haselberger PA, Hektor T, Hoefling R, Holroyd S, Lloyd Holt D, Horst JG, Ivory R, Jaureguibeitia A, Jennens M, Kavolis DC, Kock L, Konings EJM, Krepich S, Krueger DA, Lacorn M, Lassitter CL, Lee S, Li H, Liu A, Liu K, Lusiak BD, Lynch E, Mastovska K, McCleary BV, Mercier GM, Metra PL, Monti L, Moscoso CJ, Narayanan H, Parisi S, Perinello G, Phillips MM, Pyatt S, Raessler M, Reimann LM, Rimmer CA, Rodriguez A, Romano J, Salleres S, Sliwinski M, Smyth G, Stanley K, Steegmans M, Suzuki H, Swartout K, Tahiri N, Eyck RT, Torres Rodriguez MG, Van Slate J, Van Soest PJ, Vennard T, Vidal R, Vinbord Hedegaard RS, Vrasidas I, Vrasidas Y, Walford S, Wehling P, Winkler P, Winter R, Wirthwine B, Wolfe D, Wood L, Woollard DC, Yadlapalli S, Yan X, Yang J, Yang Z, Zhao G. (2018b) Standard method performance requirements (SMPRs®) 2018.002: fructans in animal food (animal feed, pet food, and ingredients). J AOAC Int101, 4:1283–1284. https://doi.org/10.5740/jaoacint.smpr2018.002

Van Ruth SM, Huisman W, Luning PA (2017) Food fraud vulnerability and its key factors. Trends Food Sci Technol 67:70–75. https://doi.org/10.1016/j.tifs.2017.06.017

Vloeberghs D, Bellens J (1996) Implementing the ISO 9000 standards in Belgium. Qual Prog 29(6):43–48

Yung WK (1997) The values of TQM in the revised ISO 9000 quality system. Int J Op Prod Manag 17(2):221–230. https://doi.org/10.1108/01443579710158078

Zaccheo A, Palmaccio E, Venable M, Locarnini-Sciaroni I, Parisi S (2017) Food hygiene and applied food microbiology in an anthropological cross cultural perspective. Springer International Publishing, Cham

Chapter 3
Quality Audits in Food Companies and the Examination of Technical Data Sheets

Abstract One of the main questions concerning raw materials and intermediates in the food industry is the definition of used ingredients by the legal viewpoint. In the ambit of quality management systems, the declaration of all possible components of a food or beverage product may concern or be correlated with some basic aspects, including the critical and reliable interpretation of technical data sheets concerning these products in the food and beverage industries by external and experienced auditors. Interestingly, the matter of technical data sheets appears to be always critical when speaking of food quality management. In general, these documents should give clear answers, and auditors should be ready to understand and analyse these information. The aim of this chapter is to give reliable advices for interested food auditors with concern to the examination of technical data sheets for all possible ingredients in the food and beverage sector.

Keywords Allergen · Audit · HACCP · Identity · Quality system · Raw material · Technical data sheet

Abbreviations

GMO   Genetically modified organism
GSFS   Global Standard for Food Safety
HACCP Hazard Analysis and Critical Control Points
HARPC Hazard Analysis and Risk-based Preventive Controls
IFS   International Featured Standard

3.1 The Importance of Technical Data Sheets in the Food Industry: Quality Management Systems

One of the main questions concerning raw materials and intermediates in the food industry is 'What is this ingredient? How should I define and declare it on labels?'

© The Author(s), under exclusive license to Springer Nature Switzerland AG 2019   39
M. Fiorino et al., *Quality Systems in the Food Industry*, Chemistry of Foods,
https://doi.org/10.1007/978-3-030-22553-7_3

In the ambit of quality management systems (Sect. 1.2), the declaration of ingredients—all possible components of a food or beverage product—may concern or be correlated with four basic aspects:

(1) The role of food additives, processing aids and chemical compounds in general (Asensio et al. 2008; Lindsay 2007; Lupien 2005; McEntire et al. 2010; Singhal et al. 1997), when speaking of traceability, possible identity loss and access in the flow diagram of processes
(2) Countermeasures against intentional adulteration (Fennema et al. 2017; Moore et al. 2012)
(3) The evaluation of technical data sheets concerning these products in the food and beverage industries by external and experienced auditors (Crossland 1997; Panisello and Quantick 2001; Powell et al. 2013)
(4) Finally, possible health risks caused by allergenic and undeclared substances, because of the demonstrated possibility and reported the occurrence of public health issues (BRC 2017a, b; Nikoleiski 2015; Stein 2015).

With specific concern to the definition of ingredients in the food and beverage ambit, quality standards generally manage this problem in the following quality areas:

(1) The 'food safety plan' according to basic principles of the Hazard Analysis and Critical Control Points (HACCP) assessment and/or the most recent approach named Hazard Analysis and Risk-based Preventive Controls (HARPC) (Baines 1999; Channaiah et al. 2017; Davidson et al. 2017; Grover et al. 2015; Parisi 2009, 2012, 2016; Schulze et al. 2008; Stilo et al. 2009)
(2) The behaviour of the food business operator when speaking of basic operations in the quality management ambit: internal audits, revealed non-conformities and consequent preventive/corrective actions (Albersmeier et al. 2009; Laux and Hurburgh 2012; Sperber 1998)
(3) The evaluation and periodical re-assessment of raw material suppliers
(4) The existence of a reliable traceability system (Mania et al. 2016, 2017, 2018a, b, c, d)
(5) The identification of possible ingredients containing allergens and/or genetically modified organisms (GMO); these information and all possible details concerning the identification of ingredients should be easily found in technical data sheets
(6) And, last but not least, the evaluation of technical data sheets concerning all possible ingredients, by the chemical viewpoint at least, even if some doubts can be expressed when speaking of peculiar and non-chemical claims (e.g. geographical origin; expression of particular ethnic groups or traditions; exclusion of certain ingredients; environmental sustainability).

Interestingly, the matter of technical data sheets appears to be critical in all above-mentioned quality areas. In fact, these important and mandatory documents have to concern:

(a) The mention of dedicated safety assessment concerning the ingredient(s) and the desired use in the food industry, sector and sub-sector
(b) Possible suggestions of use or countermeasures if the correct dosage is not properly performed
(c) The identification of raw material suppliers and producers above all (the identity producer = supplier is not often observed)
(d) Traceability information (lot explanation, declared durability, etc.
(e) The identification of all possible raw materials in the formulation of the ingredient(s), including possible allergens and GMO
(f) Possible claims when speaking of nutritional properties and ethic choices.

In general, these documents should give clear unambiguous answers to these questions. The following section is dedicated to the study of these technical data sheets with advices for interested food auditors.

The new Global Standard for Food Safety (GSFS), Issue 8, and the International Food Standard (IFS), Version 6.1, give some useful guideline when speaking of technical data sheets and related requirements. In particular, GSFS Issue 8 states that finished product specifications should at least include the following information (BRC 2018):

(a) The list of ingredients, including also possible allergens
(b) The nutritional data
(c) Some instructions for preparation and use
(d) Storage requirements
(e) Explicit durability
(f) Declared amount (as weight or volumetric capacity).

The same requirements should be met for all possible raw materials, intermediates, packaging materials, food-grade lubricants and other food contact materials. As above shown, these requisites are a minimum request, while other data are often requested at present (BRC 2018; ISO 2006).

3.2 Technical Data Sheets: Commercial Name or Brand, Producer Name, Description of the Product, Declared or Intended Use, Dosages

Figure 3.1 shows a technical data sheet (a template) concerning a hypothetical product for subsequent use into a food company. This picture can serve as a general template for the redaction of technical data sheets concerning both finished products and raw materials. It has to be noted that each utilisation of this template has to be correlated with mentioned problems concerning food additives in general (Chap. 1), adulteration concerns (Chap. 2) and the matter of allergens (Chap. 4). Briefly, the shown information should be listed as follows (the picture is not mandatory, and it can serve as a simple template):

'FOOD INGREDIENT' xx/8t

General description: **a powdered mixture of '…' ingredients**

Recommended Uses:

Recommended dosage:

Basic information

1) Country of Origin:
2) Legal Classification or Identity: (legal reference: …………………………………………)
3) Declared Durability: xx months at yy °
4) Quality and Safety Declarations:
5) Safety Risks and Advices:
6) Material Safety Data Sheets: In-hand Document

List of Ingredients:

Nutritional information per 100 g/ml of finished product

Declared Claims

Chemical data
Microbiological data
Sensorial data

Approval (Legal representative)   **Date of Issue:**

Fig. 3.1 A technical data sheet concerning a hypothetical product for subsequent use into a food company. The commercial name or brand of the product logo, and the same food company do not exist. These names and logos are for example purposes only

(1)   The commercial name or brand of the product logo (this name and brand do not exist. These names and logos are for example purposes only)
(2)   The name and address of the producer, with a 'F&BP' logo (this logo and the company do not exist. These names and logos are for example purposes only)
(3)   A general description for this product
(4)   The intended or declared use (alternative usages may be described)
(5)   Recommended dosages.

Figure 3.1 shows also the following, both mandatory and additional data:

(1)   Country of origin
(2)   Legal classification or identity
(3)   Declared durability
(4)   Quality and safety declarations
(5)   Safety risks and advices
(6)   Material safety data sheets (in-hand documents)
(7)   List of ingredients
(8)   Nutritional information
(9)   Declared claims
(10) Chemical and microbiological performances
(11) Sensorial evaluation parameters
(12) Approval and date.

These main information may be integrated with other data. A similar template should contemplate the main group of information auditors, that should be expect to see in technical data sheets. In fact, the general risk assessment and quality management procedures rely on these basic information.

References

Albersmeier F, Schulze H, Jahn G, Spiller A (2009) The reliability of third-party certification in the food chain: from checklists to risk-oriented auditing. Food Control 20(10):927–935. https://doi.org/10.1016/j.foodcont.2009.01.010

Asensio L, González I, García T, Martín R (2008) Determination of food authenticity by enzyme-linked immunosorbent assay (ELISA). Food Control 19(1):1–8. https://doi.org/10.1016/j.foodcont.2007.02.010

Baines RN (1999) Environmental & animal safety dimensions to developing food safety & quality assurance initiatives in the United Kingdom. Market rights and equity: food and agriculture standards in a shrinking world. Institute for Food & Agricultural Standards, Michigan

BRC (2017a) Global standard food safety, issue 7—An introduction to best-practice lubrication procedures in the food industry. British Retail Consortium (BRC) Global Standards, London, www.brcglobalstandards.com. Available https://www.brcgs.com/media/165616/brc-lubrication-guide-screen.pdf. Accessed 10 Apr 2019

BRC (2017b) An introduction to best-practice lubrication procedures in the food industry, issue 1, April 2017. British Retail Consortium (BRC) Global Standards, London, www.brcglobalstandards.com

BRC (2018) Global standard food safety, issue 8. Interpretation guideline. British Retail Consortium (BRC) Global Standards, London, www.brcglobalstandards.com

Channaiah LH, Michael M, Acuff JC, Phebus RK, Thippareddi H, Olewnik M, Milliken G (2017) Validation of the baking process as a kill-step for controlling Salmonella in muffins. Int J Food Microbiol 250:1–6. https://doi.org/10.1016/j.ijfoodmicro.2017.03.007

Crossland WJ (1997) HACCP and factory auditing. In: Chesworth N (ed) Food hygiene auditing. Chapman & Hall, London, pp 30–52

Davidson RK, Antunes W, Madslien EH, Belenguer J, Gerevini M, Torroba Perez T, Prugger R (2017) From food defence to food supply chain integrity. Brit Food J 119(1):52–66. https://doi.org/10.1108/BFJ-04-2016-0138

Fennema OR, Damodaran S, Parkin KL (2017) Introduction to food chemistry. In: Damodaran S, Parkin KL (eds) Fennema's food chemistry, 5th edn. CRC Press, Boca Raton, pp 1–16

Grover AK, Chopra S, Mosher GA (2015) Adoption of food safety modernization act: a six sigma approach to risk based preventive controls for small food facilities. In: Conference proceeding of the 2015 annual ATMAE conference, Pittsburgh, 11–14 November 2015. Available https://lib.dr.iastate.edu/cgi/viewcontent.cgi?article=1464&context=abe_eng_conf. Accessed 10 Apr 2019

ISO (2006) ISO 21469:2006—Safety of machinery—Lubricants with incidental product contact—Hygiene requirements. International Organization for Standardization, Geneva. Available https://www.iso.org/standard/35884.html. Accessed 11 Apr 2019

Laux CM, Hurburgh CR Jr (2012) Using quality management systems for food traceability. J Ind Technol 26(3):1–10. Available https://lib.dr.iastate.edu/cgi/viewcontent.cgi?article=1717&context=abe_eng_pubs. Accessed 11 Apr 2019

Lindsay RC (2007) Food additives. In: Parkin KL, Fennema OR (eds) Fennema's food chemistry. CRC Press, Boca Raton, pp 701–762

Lupien JR (2005) Food quality and safety: traceability and labeling. Crit Rev Food Sci Nutr 45(2):119–123. https://doi.org/10.1080/10408690490911774

Mania I, Fiorino M, Barone C, Barone S, Parisi S (2016) Traceability of packaging materials in the cheesemaking field. The EU regulatory ambit. Food Packag Bull 25(4&5):11–16

Mania I, Barone C, Pellerito A, Laganà P, Parisi S (2017) Trasparenza e Valorizzazione delle Produzioni Alimentari. 'etichettatura e la Tracciabilità di Filiera come Strumenti di Tutela delle Produzioni Alimentari. Ind Aliment 56, 581:18–22

Mania I, Delgado AM, Barone C, Parisi S (2018a) Traceability in the dairy industry in Europe. Springer International Publishing, Heidelberg, Germany. https://doi.org/10.1007/978-3-030-00446-0

Mania I, Delgado AM, Barone C, Parisi S (2018b) Food packaging and the mandatory traceability in Europe. In: Traceability in the dairy industry in Europe. Springer International Publishing, Heidelberg, Germany. https://doi.org/10.1007/978-3-030-00446-0_8

Mania I, Delgado AM, Barone C, Parisi S (2018c) The ExTra tool—A practical example of extended food traceability for cheese productions. In: Traceability in the dairy industry in Europe. Springer International Publishing, Heidelberg, Germany. https://doi.org/10.1007/978-3-030-00446-0_3

Mania I, Delgado AM, Barone C, Parisi S (2018d) Food additives for analogue cheeses and traceability: the ExTra tool. In: Traceability in the dairy industry in Europe. Springer International Publishing, Heidelberg, Germany. https://doi.org/10.1007/978-3-030-00446-0_7

McEntire JC, Arens S, Bernstein M, Bugusu B, Busta FF, Cole M, Davis A, Fisher W, Geisert S, Jensen H, Kenah B, Lloyd B, Mejia C, Miller B, Mills R, Newsome R, Osho k, Prince G, Scholl S, Sutton D, Welt B, Ohlhorst S (2010) Traceability (product tracing) in food systems: an IFT report submitted to the FDA, vol 1: technical aspects and recommendations. Compr Rev Food Sci Food Saf 9(1):92–158. https://doi.org/10.1111/j.1541-4337.2009.00097.x

Moore JC, Spink J, Lipp M (2012) Development and application of a database of food ingredient fraud and economically motivated adulteration from 1980 to 2010. J Food Sci 77(4):R118–R126. https://doi.org/10.1111/j.1750-3841.2012.02657.x

Nikoleiski D (2015) Hygienic design and cleaning as an allergen control measure. In: Flanagan S (ed) Handbook of food allergen detection and control, Woodhead Publishing series in food science, technology and nutrition, pp 89–102. Woodhead Publishing Ltd, Cambridge. https://doi.org/10.1533/9781782420217.1.89

Panisello PJ, Quantick PC (2001) Technical barriers to hazard analysis critical control point (HACCP). Food Control 12(3):165–173. https://doi.org/10.1016/S0956-7135(00)00035-9

Parisi S (2009). Intelligent packaging for the food industry. Polymer electronics—a flexible technology. Smithers Rapra Technology Ltd, Shawbury

Parisi S (2012) Food packaging and food alterations. The user-oriented approach. Smithers Rapra Technology Ltd, Shawbury

Powell DA, Erdozain S, Dodd C, Costa R, Morley K, Chapman BJ (2013) Audits and inspections are never enough: a critique to enhance food safety. Food Control 30(2):686–691. https://doi.org/10.1016/j.foodcont.2012.07.044

Schulze H, Albersmeier F, Gawron JC, Spiller A, Theuvsen L (2008) Heterogeneity in the evaluation of quality assurance systems: the International Food Standard (IFS) in European agribusiness. Int Food Agribus Manag Rev 119(1):52–66. https://doi.org/10.1108/BFJ-04-2016-0138

Singhal RS, Kulkarni PK, Reg DV (eds) (1997) Handbook of indices of food quality and authenticity. Woodhead Publishing Ltd, Cambridge

Sperber WH (1998) Auditing and verification of food safety and HACCP. Food Control 9(2–3):157–162. https://doi.org/10.1016/S0956-7135(97)00068-6

Stein K (2015) Effective allergen management practices to reduce allergens in food. In Handbook of food allergen detection and control. In: Flanagan S (ed) Handbook of food allergen detection and control, Woodhead Publishing series in food science, technology and nutrition, pp 103–131. Woodhead Publishing Ltd, Cambridge. https://doi.org/10.1533/9781782420217.1.103

Stilo A, Parisi S, Delia S, Anastasi F, Bruno G, Laganà P (2009) Food security in Europe: comparison between the "Hygiene Package" and the British Retail Consortium (BRC) & International Food Standard (IFS) protocols. Ann Ig 21(4):387–401

Chapter 4
Allergen Risks and the Use of Certified Lubricants in the Modern Food Industry

Abstract This chapter describes the certification of food-grade additives, raw materials and non-food materials such as lubricants in the food industry, especially with regard to contained allergenic substances. The problem of allergens should be discussed in the ambit of quality management systems because the definition of allergen concerns potentially all food contact ingredients and materials in the food or beverage industry. By the viewpoint of quality managers, the risk of declared and undeclared allergens is critical and should also take into account 'hidden' components which could enter the food processing chain, such as certain lubricants used for food processing equipment. These topics are well considered and discussed when speaking of the most known and used standards in the ambit of food certification, including the Global Standard for Food Safety by the British Retail Consortium and the International Featured Standard Food. This chapter discusses general requests for allergens by the quality management viewpoint and the recommended management of non-food materials such as lubricants.

Keywords Allergen · Cross-contamination · Food-grade lubricant · GMO · GSFS · IFS · Processing flow

Abbreviations

BRC	British Retail Consortium
EU	European Union
FSSC	Food Safety System Certification
GFSI	Global Food Safety Initiative
GSFS	Global Standard for Food Safety
HACCP	Hazard Analysis and Critical Control Points
IFS	International Featured Standard
ISO	International Organization for Standardization
NSF	National Sanitation Foundation
SQF	Safe Quality Food

© The Author(s), under exclusive license to Springer Nature Switzerland AG 2019
M. Fiorino et al., *Quality Systems in the Food Industry*, Chemistry of Foods,
https://doi.org/10.1007/978-3-030-22553-7_4

4.1 Food Allergens and Unknown Sources

Why the problem of allergens should be discussed in the ambit of quality management systems?

Basically, the definition of allergenic substance (Sect. 4.2) concerns potentially all—both main and minor—ingredients of the food or beverage product. By the viewpoint of quality managers, the risk of declared and undeclared allergens (BRC 2017a; Nikoleiski 2015; Stein 2015) is critical and should also take into account 'hidden' components which could enter the food processing chain, such as certain lubricants used for food processing equipment. These topics are well considered and discussed when speaking of the most known and used standards in the ambit of food certification, including (BRC 2018a; FSSC 2017; IFS 2017; ISO 2018; SQF 2017):

- The Global Standard for Food Safety (GSFS) by the British Retail Consortium (BRC)
- The International Featured Standard (IFS) Food, by the IFS
- The ISO 22000 norm, by the International Organization for Standardization (ISO)
- The Food Safety System Certification (FSSC) 22000
- The Safe Quality Food (SQF) by the SQF Institute, USA.

The management of allergen risks—as a matter of public health and food safety—and the use of peculiar lubricants should be discussed in the following areas concerning quality management standards:

(a) The commitment of senior management because allergens are one of the most critical areas in food industries. Moreover, legal requirements such as the Food Safety Modernization Act (FSMA) in the USA require a well-documented and implemented preventive approach for this danger (Biresaw 2014)
(b) The search for continuous improvement tends to the minimisation of the presence of allergens in the food chain, even if 'zero presence' or 'zero detection' is not possible (Popping et al. 2018)
(c) The elaboration of the 'food safety plan' (Stein 2015)
(d) Demonstrable results obtained by the organisation in terms of internal audits, possible non-conformities and related preventive/corrective actions
(e) The continuous surveillance concerning the whole supply food chain (materials and related suppliers, including also all possible non-food materials or accessory components such as lubricants)
(f) The existence of a good traceability system (Mania et al. 2017, 2018a, b, c)
(g) The creation, implementation and continuous amelioration of layout designs, flow diagrams, processing operations and segregation
(h) Good operative procedures concerning cleaning and sanitisation operations (Cramer 2016; Lelieveld et al. 2003; Marriott et al. 2018; Mortimore and Wallace 2013)
(i) Dedicated training and attention to safety culture (Overbosch and Blanchard 2014; Sinha 2014; Soman and Raman 2016; Trienekens and Zuurbier 2008; Van der Spiegel et al. 2003; Zaccheo et al. 2017).

How the allergen risk should be managed? With exclusive relation (for now) to edible components, the following section discusses general requests for allergens by the quality management viewpoint. On the other side, non-food materials such as lubricants are discussed in Sect. 4.3.

4.2 Food Allergens—Quality Management Requirements

As a little premise, it should be noted that allergens can be identified as known food components 'which causes physiological reactions due to an immunological response' according to the Global Standard for Food Safety (GSFS) Issue 8 (BRC 2018b).

On these bases, the safe management of allergens is stated clearly when speaking of fundamental requirements and prerequisite programmes (clause 2.2.1) with the aim of avoiding dangerous cross-contamination episodes (BRC 2018b). This request concerns not only the prediction of possible allergens along the supply chain, but also the elimination of risks into sites which explicitly use different raw materials including allergenic substances. Because of the possible and accidental cross-contamination, adequate cleaning procedures and production methods (such as temporally separated productions concerning high-risk allergenic productions and no-risk allergenic productions) have to be considered and put in place. Also, the analytical verification for effective removal of allergens is required with another requisite: cleaning-in-place systems, if existing, should not be the cause of the reintroduction of allergenic substances (clauses 2.7.3, 4.11.1, 4.11.3 and 4.11.7.1).

Another important point (Chap. 2) concerns the monitoring surveillance of raw material suppliers. Clause 3.5.1.1 of GSFS clearly requires that the evaluation of all suppliers has to take into account also potential menaces for allergen contamination. This evaluation, based on technical data sheets (Chap. 3) and other information, has to take also into account the problem of spatial or time segregation for possible and declared allergenic substances. Consequently, raw material suppliers are requested to give all possible information concerning their supplied products (GSFS Issue 8, clause 3.5.2.1). In this ambit, GSFS requires the existence of allergen results on the analytical ground, instead of simple sheets or specific declarations, as required by GSFS Issue 8, clause 3.6.2 (BRC 2018b).

In relation to the management of production, as already clarified in Chap. 1, the description of the product has to rely also on the known composition and explicitly in relation to allergens (clause 2.3.1). Also, the design of layouts and flow diagrams has to take into account cross-contamination risks with explicit reference to allergen controls and the position of potential allergens on site (GSFS Issue 8, clauses 2.5.1 and 4.3.1). Also, transport should take into account a specified 'allergen risk' because of the possible existence of different food typologies in the same truck (GSFS Issue 8, clause 4.16.5)!

The same design considerations are declared when speaking of the food safety or 'Hazard Analysis and Critical Control Points' (HACCP) plan (clause 2.7.1). With concern to allergens, the problem is also dependent on the peculiar type and class of allergenic substance or material, as stated in Sect. 5.3 of GSFS Issue 8, and in relation to 'offline' or reworking materials (if these materials may contain allergens) and their re-access into a processing flow (GSFS Issue 8, clause 5.3.5) (BRC 2018b). It has been also requested that food catering facilities and vending machines cannot be the cause of possible and unintentional access of allergenic substances in production and manipulation areas. This request concerns the creation and implementation of adequate behaviour procedures with reference to operators and external visitors; also, some analytical control has to be put in place (GSFS Issue 8, clauses 4.8.8 and 5.3.4).

Another important—and apparently 'accessory'—requirement concerns packaging materials because of the needed control on labelled information. In other words, only checked and authorised packaging materials can be used into food industries, and they must comply with standards and existing legislation, with specific relation to the obligatory mention of allergens, if any, and possible claims such as 'allergen-free' phrases (GSFS Issue 8, clauses 3.5.2.2, 5.2.1, and 5.2.3).

Finally, what should be mentioned allergens? In general, the main list is found in the Codex General Standard for the Labelling of Prepackaged Foods (CODEX STAN 1-1985, paragraph 4.2.1.4). However, several countries can have different lists: the European Union (EU), Australia, Japan and the USA have their own lists (BRC 2018b). It has to be remembered that the analysis concerning allergens has to be carried out on raw materials with explicit reference to the status of all ingredients (Sect. 1.3). The important point is that risk assessment has to consider also all ingredients of compound ingredients, if any, used in the food industry. This 'second-level' analysis of ingredients manufactured by other producers is extremely important and is strongly requested by GSFS Issue 8, clause 5.3.1 (BRC 2018b). In this ambit, resulting from the complete vulnerability risk assessment, a complete and exhaustive and updated list of all handled allergens, differentiates also with consideration to their physical state (powdered, liquid, fatty material, etc.) has to be constantly maintained by the food and business operator (clause 5.3.2). The physical nature of allergens is particularly important, especially in relation to possible aerosolised suspensions or strong adherence to surfaces into the site. In the broad ambit of regulatory, a good example of allergen lists can be obtained according to the Regulation (EU) No 1169/2011 as follows (European Parliament and Council 2011):

(a) Gluten-containing cereals (i.e. wheat, rye, barley, oats, spelt, *Khorasan* or their hybridised strains) and products thereof
(b) Crustaceans and products thereof
(c) Eggs and products thereof
(d) Fish and products thereof
(e) Peanuts and products thereof
(f) Soya beans and products thereof
(g) Milk and products thereof (including lactose)
(h) Nuts, i.e. almond (*Amygdalus communis L.*), hazelnut (*Corylus avellana*), walnut (*Juglans regia*), cashew (*Anacardium occidentale*), pecan nut (*Carya illinoiesis (Wangenh.) K. Koch*), Brazil nut (*Bertholletia excelsa*), pistachio nut

(*Pistacia vera*), macadamia nut, Queensland nut (*Macadamia ternifolia*) and products thereof
(i)   Celery and products thereof
(j)   Lupin and products thereof
(k)   Molluscs and products thereof
(l)   Mustard and products thereof
(m) Sesame seeds and products thereof
(n)   Sulphur dioxide and sulphites at concentrations of
(o)   more than 10 mg/kg or 10 mg/litre expressed as sulphur dioxide.

A possible allergen list for a general food and business operator is shown in Table 4.1. This 'blank' list shows all known allergens and possible access points (formulation, physical presence on site, possibility of line contamination because of concomitant processing flows, cross-contamination). This list may be useful enough when speaking of the correct management of allergenic substances into food industries. The presence of certain non-food materials with allergenic properties should be still discussed in Sect. 4.3.

4.3   Food-Grade Lubricants—Are They Possible Allergens?

Why should food-grade lubricants be considered when speaking of allergen risks in the food production ambit?

According to GSFS Issue 8, clause 4.7.5, the problem of lubricating oils is mentioned generically when speaking of 'materials and parts used for equipment and plant maintenance'. This requirement states clearly that only food-grade materials with 'appropriate grade or quality' can be used, with the aim of avoiding risks because of direct or indirect contacts with raw materials, intermediate and final products, including also separated parts such as packaging materials. The reference to allergens is made when speaking of known 'allergen status' (BRC 2018b). For this reason, only food-grade lubricants with known allergenic status can be allowed: the supplier of these products—and other non-lubricant products—has to give evidence of the absence or presence of allergens with dedicated data sheets (and material safety data sheets) or declarations, although a specific statement mentioning the ISO 21469:2006 norm for lubricants can be surely issued (and accepted) (ISO 2006; Raab 2002).

On the other hand, the role of 'machinery lubricants' is mentioned clearly when speaking of potential hazards (GSFS Issue 8, clause 2.7.1), surveillance actions for gases and steam (clause 4.5.3, because lubricants can be dispersed and aerosolised) and safe use in general as mentioned by GSFS Issue 8, clause 4.9.1.1 (BRC 2018b). The National Sanitation Foundation (NSF) International has proposed its own classification of food-grade lubricants (BRC 2017b).[1] On the other hand, the ISO

---

[1] This information can be available at the following website: http://info.nsf.org/USDA/psnclistings.asp.

Table 4.1 A possible allergen list for a general food and business operator. This 'blank' list shows all possible allergens and possible access points (formulation, physical presence on site, possibility of line contamination because of concomitant processing flows, cross-contamination)

Allergen list—HACCP plan

These allergens are present in the formulation or due to cross-contamination episodes

	Formula?	On site?	Processing flow?	Cross-contamination?
Gluten-containing cereals (i.e. wheat, rye, barley, oats, spelt, *Khorasan* or their hybridised strains) and products thereof				
Crustaceans and products thereof				
Eggs and products thereof				
Fish and products thereof				
Peanuts and products thereof				
Soya beans and products thereof				
Milk and products thereof (including lactose)				
Nuts, i.e. almond (*Amygdalus communis* L.), hazelnut (*Corylus avellana*), walnut (*Juglans regia*), cashew (*Anacardium occidentale*), pecan nut (*Carya illinoiesis* (*Wangenh.*) *K. Koch*), Brazil nut (*Bertholletia excelsa*), pistachio nut (*Pistacia vera*), macadamia nut, Queensland nut (*Macadamia ternifolia*) and products thereof				
Celery and products thereof				
Lupin and products thereof				
Molluscs and products thereof				
Mustard and products thereof				
Sesame seeds and products thereof				
Sulphur dioxide and sulphites at concentrations of more than 10 mg/kg or 10 mg/litre expressed as sulphur dioxide				

21469:2006 norm has recently considered this matter with reference to the following factors (BRC 2017b):

(a) The formulation of the lubricant
(b) The production process in terms of good manufacturing practices, similarly to the approach used for food contact packaging materials and objects (Parisi 2012)
(c) Traceability procedures concerning components of the lubricant
(d) Safety requirements concerning the lubricant.

Consequently, the allergen status of lubricants—and their food-grade status in general—has to be known. Here, you have, in conclusion, a little list (by the quality auditor's viewpoint) containing several minimum requirements which should be met when speaking of food-grade lubricants in a safe and allergen-free environment (BRC 2017b):

(1) Their use is defined as acceptable when:

   (1.1) Incidental contact in food processing areas is possible (the NSF classification is NSF H1), or
   (1.2) Incidental contact in food processing areas is possible and this lubricant is used as heat transfer medium (the NSF classification is NSF HT1), or
   (1.3) The use aims at the prevention or food adhesion during processing. In these situations, allowed food productions concern the use of hard surfaces in contact with meat and/or poultry products. The NSF classification is NSF 3H.

(2) These substances are stored and segregated in different warehouses or areas for food-grade lubricants only. In other terms, these areas cannot contain non-food-grade lubricants (or discarded/wasted oils and suspensions).
(3) The removal of used lubricants is carried out with the aim of eliminating excess oils; also, wasted oils have to be removed and eliminated separately from other products, without cross-contamination risks.
(4) Their allergen status is well known. Should these materials be allergenic, their use has to be correctly managed (Sects. 4.1 and 4.2).
(5) Their use has to be approved if peculiar products such as halal or kosher foods are prepared.

References

Biresaw G (2014) Environmentally friendly lubricant-development programs at USDA. In: ASTM D02.12 Subcommittee on Environmental Standards of Lubricants, STP1575—Environmentally Considerate Lubricants. ASTM International, West Conshohocken. https://doi.org/10.1520/stp157520130172

BRC (2017a) Global Standard Food Safety Issue 7—an introduction to best-practice lubrication procedures in the food industry. British Retail Consortium (BRC) Global Standards, London. www.brcglobalstandards.com. Available https://www.brcgs.com/media/165616/brc-lubrication-guide-screen.pdf. Accessed 10 Apr 2019

BRC (2017b) An introduction to best-practice lubrication procedures in the food industry, Issue 1—April 2017. British Retail Consortium (BRC) Global Standards, London. www.brcglobalstandards.com

BRC (2018a) Global Standard Food Safety, Issue 8. British Retail Consortium (BRC) Global Standards, London. www.brcglobalstandards.com

BRC (2018b) Global Standard Food Safety, Issue 8. Interpretation Guideline. British Retail Consortium (BRC) Global Standards, London. www.brcglobalstandards.com

Cramer MM (2016) Food plant sanitation: design, maintenance, and good manufacturing practices, 2nd edn. CRC Press, Boca Raton, London, New York

European Parliament and Council (2011) Regulation (EU) No 1169/2011 of the European Parliament and of the Council of 25 October 2011 on the provision of food information to consumers, amending Regulations (EC) No 1924/2006 and (EC) No 1925/2006 of the European Parliament and of the Council, and repealing Commission Directive 87/250/EEC, Council Directive 90/496/EEC, Commission Directive 1999/10/EC, Directive 2000/13/EC of the European Parliament and of the Council, Commission Directives 2002/67/EC and 2008/5/EC and Commission Regulation (EC) No 608/2004. Off J Eur Union L 304(18):18–63

FSSC (2017) FSSC 22000 version 4.1. Foundation FSSC 22000, Gorinchem. http://www.fssc22000.com

IFS (2017) IFS Food—standard for auditing quality and food safety of food products, Version 6.1 November 2017. International Featured Standards (IFS) Management GmbH, Berlin

ISO (2006) ISO 21469:2006—Safety of machinery-lubricants with incidental product contact—hygiene requirements. International Organization for Standardization, Geneva. Available https://www.iso.org/standard/35884.html. Accessed 11 Apr 2019

ISO (2018) ISO 22000:2018—Food safety management systems—requirements for any organization in the food chain. ISO TC 34/SC 17 (Management systems for food safety). International Organization for Standardization, Geneva. Available https://www.iso.org/standard/65464.html

Lelieveld HL, Mostert MA, Holah J, White B (eds) (2003) Hygiene in food processing: principles and practice. Woodhead Publishing Ltd, Cambridge

Mania I, Barone C, Pellerito A, Laganà P, Parisi S (2017) Trasparenza e Valorizzazione delle Produzioni Alimentari. 'etichettatura e la Tracciabilità di Filiera come Strumenti di Tutela delle Produzioni Alimentari. Ind Aliment 56(581):18–22

Mania I, Delgado AM, Barone C, Parisi S (2018a) Food packaging and the mandatory traceability in Europe. In: Traceability in the dairy industry in Europe. Springer International Publishing, Heidelberg. https://doi.org/10.1007/978-3-030-00446-0_8

Mania I, Delgado AM, Barone C, Parisi S (2018b) The ExTra Tool—a practical example of extended food traceability for cheese productions. In: Traceability in the dairy industry in Europe. Springer International Publishing, Heidelberg. https://doi.org/10.1007/978-3-030-00446-0_3

Mania I, Delgado AM, Barone C, Parisi S (2018c) Food additives for analogue cheeses and traceability: The ExTra Tool. In: Traceability in the dairy industry in Europe. Springer International Publishing, Heidelberg. https://doi.org/10.1007/978-3-030-00446-0_7

Marriott NG, Schilling MW, Gravani RB (2018) Principles of food sanitation, Food Science Text Series, vol 22. Springer International Publishing AG, Cham. https://doi.org/10.1007/978-3-319-67166-6

Mortimore S, Wallace C (2013) Prerequisites for food safety: PRPs and operational PRPs. In: Mortimore S, Wallace C (eds) HACCP. Springer, Boston, pp 113–154. https://doi.org/10.1007/978-1-4614-5028-3_4

Nikoleiski D (2015) Hygienic design and cleaning as an allergen control measure. In: Flanagan S (ed) Handbook of food allergen detection and control, Woodhead Publishing Series in Food Science, Technology and Nutrition. Woodhead Publishing Ltd, Cambridge, pp 89–102. https://doi.org/10.1533/9781782420217.1.89

Overbosch P, Blanchard S (2014) Principles and systems for quality and food safety management. In: Motarjemi Y, Lelieveld H (eds) Food safety management. Academic Press, London, Waltham, San Diego, pp 537–558. https://doi.org/10.1016/b978-0-12-381504-0.00022-6

Parisi S (2012) Food packaging and food alterations. The user-oriented approach. Smithers Rapra Technology Ltd, Shawbury

Popping B, Allred L, Bourdichon F, Brunner K, Diaz-Amigo C, Galan-Malo P, Lacorn M, North J, Parisi S, Rogers A, Sealy-Voyksner J, Thompson T, Yeung J (2018) Stakeholders' guidance document for consumer analytical devices with a focus on gluten and food allergens. J AOAC 101(1):1–5. https://doi.org/10.5740/jaoacint.17-0425

Raab MJ (2002) Food-grade lubricants: A new world order Assuring food safety in food processing: the future regulatory environment for food-grade lubricants. Tribol Lubr Technol 58(2):16–20

Sinha S (2014) Creating a culture of compliance across your supply chain. Qual Dig. https://www.qualitydigest.com. Available https://www.qualitydigest.com/inside/quality-insider-article/creating-culture-compliance-across-your-supply-chain.html. Accessed 11 Apr 2019

Soman R, Raman M (2016) HACCP system–hazard analysis and assessment, based on ISO 22000: 2005 methodology. Food Control 69:191–195. https://doi.org/10.1016/j.foodcont.2016.05.001

SQF (2017) SQF quality code, 8th edn. Safe Quality Food (SQF) Institute, Arlington. Available https://www.sqfi.com/what-is-the-sqf-program/sqf-quality-program/. Accessed 10 Apr 2019

Stein K (2015) Effective allergen management practices to reduce allergens in food. In: Flanagan S (ed) Handbook of food allergen detection and control, Woodhead Publishing Series in Food Science, Technology and Nutrition. Woodhead Publishing Ltd, Cambridge, pp 103–131. https://doi.org/10.1533/9781782420217.1.103

Trienekens J, Zuurbier P (2008) Quality and safety standards in the food industry, developments and challenges. Int J Prod Econ 113(1):107–122. https://doi.org/10.1016/j.ijpe.2007.02.050

Van der Spiegel M, Luning PA, Ziggers GW, Jongen WMF (2003) Towards a conceptual model to measure effectiveness of food quality systems. Trends Food Sci Technol 14(10):424–431. https://doi.org/10.1016/S0924-2244(03)00058-X

Zaccheo A, Palmaccio E, Venable M, Locarnini-Sciaroni I, Parisi S (2017) Food hygiene and applied food microbiology in an anthropological cross cultural perspective. Springer International Publishing, Cham

Correction to: Quality Systems in the Food Industry

Correction to:
M. Fiorino et al., *Quality Systems in the Food Industry*,
Chemistry of Foods,
https://doi.org/10.1007/978-3-030-22553-7

The original version of the book was published with incorrect affiliation of the book author "Arpan Bhagat". The book has been updated with this change.

The updated version of this chapter can be found at
https://doi.org/10.1007/978-3-030-22553-7

© The Author(s), under exclusive license to Springer Nature Switzerland AG 2019   C1
M. Fiorino et al., *Quality Systems in the Food Industry*, Chemistry of Foods,
https://doi.org/10.1007/978-3-030-22553-7_5

Printed in the United States
By Bookmasters